BRITISH GEOLOGICAL SURVEY

R. A. ELLISON and
R. D. LAKE

CONTRIBUTORS

Stratigraphy
C. R. Bristow

Economic geology
P. M. Hopson

Petrology
R. J. Merriman

Geology of the country around Braintree

Memoir for 1:50 000 geological sheet 223 (England and Wales)

Natural Environment Research Council

LONDON: HER MAJESTY'S STATIONERY OFFICE   1986

iv

© *Crown copyright 1986*

First published 1986

ISBN 0 11 884393 1

Bibliographical reference

ELLISON, R. A. and LAKE, R. D. 1986. Geology of the country around Braintree. *Mem. Br. Geol. Surv.*, Sheet 223, 69 pp.

Authors

R. A. Ellison, BSc and R. D. Lake, MA
British Geological Survey, Keyworth, Nottingham NG12 5GG

Contributors

P. M. Hopson, BSc and R. J. Merriman, BSc
British Geological Survey, Keyworth

C. R. Bristow, BSc, PhD
British Geological Survey
St Just, 30 Pennsylvania Road
Exeter EX4 6BX

Other publications of the Survey dealing with this district and adjoining districts

BOOKS

British Regional Geology
London and the Thames Valley (3rd Edition)

Memoir
Geology of the country around Chelmsford, Sheet 241

Well Catalogue
Records of wells in the area of New Series one-inch (geological) Great Dunmow (222) and Braintree (223) sheets

MAPS

1:625 000
Solid geology (South sheet)
Quaternary geology (South sheet)
Aeromagnetic map (South sheet)

1:50 000 and 1:63 360 (Solid and Drift)
Sheet 205   Saffron Walden (1952)
Sheet 206   Sudbury (1950)
Sheet 207   Ipswich (1965)
Sheet 240   Epping (1981)
Sheet 241   Chelmsford (1975)

Printed for Her Majesty's Stationery Office by W. & J. Linney Ltd.
Dd 0238611 C20 11/86 49913

Geology of the country around Braintree

The landscape of the district described in this memoir is dominated by a gently undulating plateau of drift, much of it a legacy of the Anglian ice-sheet which once covered all but the south-east of the district. The drift largely masks the solid rocks which are mainly London Clay, though the Upper Chalk and older Tertiary rocks form rock-head in the north-west; there are patches of shelly Crag in the Stour valley.

Along the deeper valleys dissecting the plateau and beyond the limit of the Anglian ice, a varied drift sequence is exposed. At or near its base there are fluvial gravels laid down by the Thames when it followed a much more northerly course than it does today. Above these come the gravels, tills and silts of the Anglian glaciation. Near Marks Tey, close to the limit of the ice, lake deposits provide perhaps the most complete British record of the subsequent Hoxnian interglacial. Comparatively thin fluvial sands and gravels together with some peat and solifluction deposits, complete the geological record.

Plate 1   Poorly sorted Glacial Sand and Gravel with a lens of well sorted sand (centre right), Stanway

CONTENTS

FIGURES

PLATES

TABLES

PREFACE

This memoir describes the geology of the district covered by the Braintree (223) New Series Sheet of the 1:50 000 Geological Map of England and Wales. The original geological survey of sheets 47 and 48NW and SW, on the scale of one inch to one mile, was carried out between 1881 and 1884 by Messrs F. J. Bennett, W. H. Dalton and W. H. Penning under the supervision of W. Whitaker. A small area in the southern part of the present sheet was surveyed on the scale of six inches to one mile by Dr C. R. Bristow and Mr R. D. Lake in 1966–69; the remainder of the district was surveyed between 1971 and 1977 by Dr R. Allender, Mr J. D. Ambrose, Dr C. R. Bristow, Dr F. C. Cox, Mr R. A. Ellison, Mr M. J. Heath, Mr R. D. Lake, Mr S. R. Mills and Dr B. S. P. Moorlock under the supervision of Mr S. C. A. Holmes and Dr W. A. Read as successive District Geologists. The new 1:50 000 map was published in 1982.

Much of this memoir has been written by Mr R. A. Ellison, but Mr R. D. Lake has been largely responsible for its compilation. In addition to the various contributions by the surveyors listed above, Mr R. J. Merriman has provided mineralogical and petrological information on the Tertiary deposits. Mr C. Turner of the Botany School, Cambridge, took part in invaluable discussions on the Quaternary deposits at Marks Tey and commented on the BGS boreholes. Mr P. M. Hopson has provided a summary of the sand and gravel resources of the district. The memoir has been edited by Mr W. B. Evans and Dr B. N. Fletcher.

Grateful acknowledgement is made to the organisations that have generously supplied borehole records and to the landowners who kindly allowed access to their ground.

G. INNES LUMSDEN, FRSE
Director

British Geological Survey
Keyworth
Nottingham   NG12 5GG

28 February 1986

LIST OF SIX-INCH MAPS

The following is a list of the six-inch National Grid geological quarter sheets which are included, wholly or in part, within the 1:50 000 Braintree (223) Geological Sheet, with the names of the surveying officers and the date of the survey of each map; the officers are R. Allender, J. D. Ambrose, C. R. Bristow, F. C. Cox, R. A. Ellison, M. J. Heath, R. D. Lake, S. R. Mills and B. S. P. Moorlock.

Manuscript copies of these maps have been deposited for public reference in the library of the British Geological Survey at Keyworth. They contain more detail than appears on the 1:50 000 map.

TL 61 NE	Ford End Lake	1969
TL 62 NE	Bran End Mills	1975
TL 62 SE	Felsted Mills	1975
TL 63 NE	Cornish Hall End Moorlock	1976
TL 63 SE	Finchingfield Moorlock	1976
TL 71 NW	Great Leighs Lake	1969
TL 71 NE	White Notley Bristow	1966–67
TL 72 NW	Shalford Ambrose, Allender and Lake	1972–76
TL 72 NE	Bocking Ambrose, Allender and Lake	1971–76
TL 72 SW	Rayne Bristow, Cox, Ellison and Lake	1971–77
TL 72 SE	Braintree Allender and Lake	1971–76
TL 73 NW	Stambourne Mills	1974
TL 73 NE	Castle Hedingham Heath	1975
TL 73 SW	Wethersfield Mills	1974–75

TL 73 SE	Sible Hedingham Heath	1975
TL 81 NW	Silver End Bristow	1967
TL 81 NE	Kelvedon Bristow	1967
TL 82 NW	Greenstead Green Heath	1974
TL 82 NE	Earls Colne Heath	1973
TL 82 SW	Bradwell Heath	1974
TL 82 SE	Coggeshall Mills	1973
TL 83 NW	Gestingthorpe Heath	1974
TL 83 NE	Lamarsh Mills	1974
TL 83 SW	Halstead Heath	1974
TL 83 SE	Pebmarsh Mills	1974
TL 91 NW	Layer Breton Bristow	1967
TL 91 NE	Peldon Bristow	1967
TL 92 NW	Fordham Ellison and Heath	1973–75
TL 92 NE	West Bergholt Ellison	1975
TL 92 SW	Marks Tey Ellison and Mills	1974–75
TL 92 SE	Stanway Ellison	1975
TL 93 NW	Assington Mills	1975
TL 93 NE	Stoke by Nayland Mills and Ellison	1974–77
TL 93 SW	Wormingford Ellison	1974–75
TL 93 SE	Great Horkesley Ellison	1974–75

NOTES

The word 'district' used in this memoir means the area included in the 1:50 000 Geological Sheet 223 (Braintree).

National Grid references are given in square brackets throughout the memoir. All lie within the 100 km square TL.

Letters preceding specimen numbers refer to Survey collections as follows:

E   National Sliced Rock collection

X   Powder films in the data bank

Numbers preceded by A refer to photographs in the Survey's collection.

The authorship of fossil species is given in the Index of fossils.

CHAPTER 1

Introduction

LOCATION AND PHYSICAL FEATURES

This memoir describes the district covered by the Braintree Sheet (New Series 223) of the 1:50 000 Geological Map of England and Wales. The district, which lies some 60 km north-east of London, is mainly in Essex, but includes a small part of Suffolk in the north-east, north of the River Stour. It lies on the northern edge of the London Basin and, except for a small tract of Chalk in the north-west, it is wholly underlain by Tertiary strata. There is an extensive cover of glacial and post-glacial deposits. The Lower London Tertiary beds crop out only in parts of the north-west of the district but they are present beneath drift deposits in a strip from north of Wethersfield to Wickham St Paul and thence into the River Stour valley. The London Clay is present at outcrop or beneath drift over the greater part of the district. It crops out mainly in the valleys but also forms a ridge near Birch Green in the south-east.

The glacial deposits comprise three main elements: Boulder Clay, Glacial Sand and Gravel, and Kesgrave Sands and Gravels. With the exception of the ground east of Stanway, between the River Colne and Roman River, where Glacial Sand and Gravel occurs, Boulder Clay forms the plateau which dominates the topography of the district. Both deposits are underlain by Kesgrave Sands and Gravels which crop out extensively on the valley sides. Post-glacial deposits are mainly confined to the river valleys and are represented by spreads of terrace gravel in the valleys of the Stour, Colne and Blackwater. Small areas of interglacial lake deposits occur in the Pods Brook valley near Braintree and in the Roman River valley at Marks Tey.

The district is drained by five major valleys, those of the Stour, Colne, Roman River, Pant–Blackwater and Brain, and their tributaries (see Figure 1). The plateau falls steadily from 107 m OD in the north-west to around 33 m in the south-east. A distinct ridge rising to 52 m in the extreme south-east marks the northern limit of the Danbury–Tiptree ridge which is a prominent physical feature in the adjoining Chelmsford (241) district to the south. The rivers drain in an easterly or south-easterly direction except for the River Blackwater which turns at Feering to run south-west through Kelvedon. The lowest part of the district comprises the flood plains of the River Stour, Roman River and River Colne which, along the eastern margin, lie at around 10 m above sea-level.

The district is dominantly agricultural, and arable farming is practised on most of the land. The Boulder Clay provides a heavy clay soil and the removal of hedges and thickets over much of its outcrop to accommodate mechanised farming techniques has resulted in a monotonous landscape. Both the London Clay and Glacial Sand and Gravel are capable of supporting arable crops, but the heavy London Clay soil needs to be well drained while the sandy and gravelly soils give lower yields: most of the woodland and scrub lie on the latter.

The rivers have played a valuable role in the economic history of the area. Particularly important was the River Colne, which in the 18th and 19th centuries supported 13 working water mills between Sible Hedingham and West Bergholt. Both the Colne and the Roman River were formerly navigable as far as Colchester. The construction of a flight of locks in 1706 made the River Stour navigable from its estuary to Sudbury, 5 km north of the district. The locks have long since fallen into disrepair and many have been replaced by flood control sluices in the last decade.

Although only the outskirts of Colchester fall within the present district, the town is worthy of note as the oldest documented town in England. Braintree, Coggeshall, Castle Hedingham, Sible Hedingham and Kelvedon are the other chief centres of population. They grew as a result of the 15th and 16th century woollen industry; the timber-framed buildings dating from this period impart to the towns much of their character. Silk manufacturing started near Braintree in 1816 and more recently light industries have been attracted to Braintree, Halstead and Sible Hedingham.

HISTORY OF RESEARCH

The geological succession in this district, first established by Prestwich (1847), was classified and described during the early work of the Geological Survey which began in the 1860s. The Old Series geological One-inch drift and solid editions of sheets 48 SW and 48 NW covering the Braintree district were surveyed on the one-inch scale and were published between 1881 and 1884. The structure and stratigraphy they portray has altered only in detail as a result of subsequent researches. The present survey is the first geological mapping on the six-inch scale in this district. The brief history of research that follows embraces major works which have affected the elucidation of the stratigraphy of the district, whilst references to more specialised detailed studies appear in the relevant general accounts of this memoir.

The first systematic research into the Tertiary deposits (now classified as Palaeocene and Eocene) was undertaken by Prestwich (1847, 1850, 1852, 1854). His work was modified and enlarged on by Whitaker (1866, 1872) who, with his colleagues Penning, Dalton and Bennett, produced the first systematic Geological Survey maps of the district. In his memoir describing the London Basin, Whitaker (1872) referred briefly to this district in his descriptions of Thanet Beds, Woolwich and Reading Beds, Oldhaven Beds and London Clay. This was followed by a more detailed memoir describing Sheet 47 (Whitaker, Penning, Dalton and Bennett, 1878) and later by a memoir on the south-western part of Sheet 48 (Dalton, 1880). The latter covers the area around Colchester and the eastern part of the Braintree district. Borehole records were included in the early Survey memoirs and have been updated by Whitaker and Thresh (1916) and, more recently, in a complete inventory for Sheet 223 (Sayer

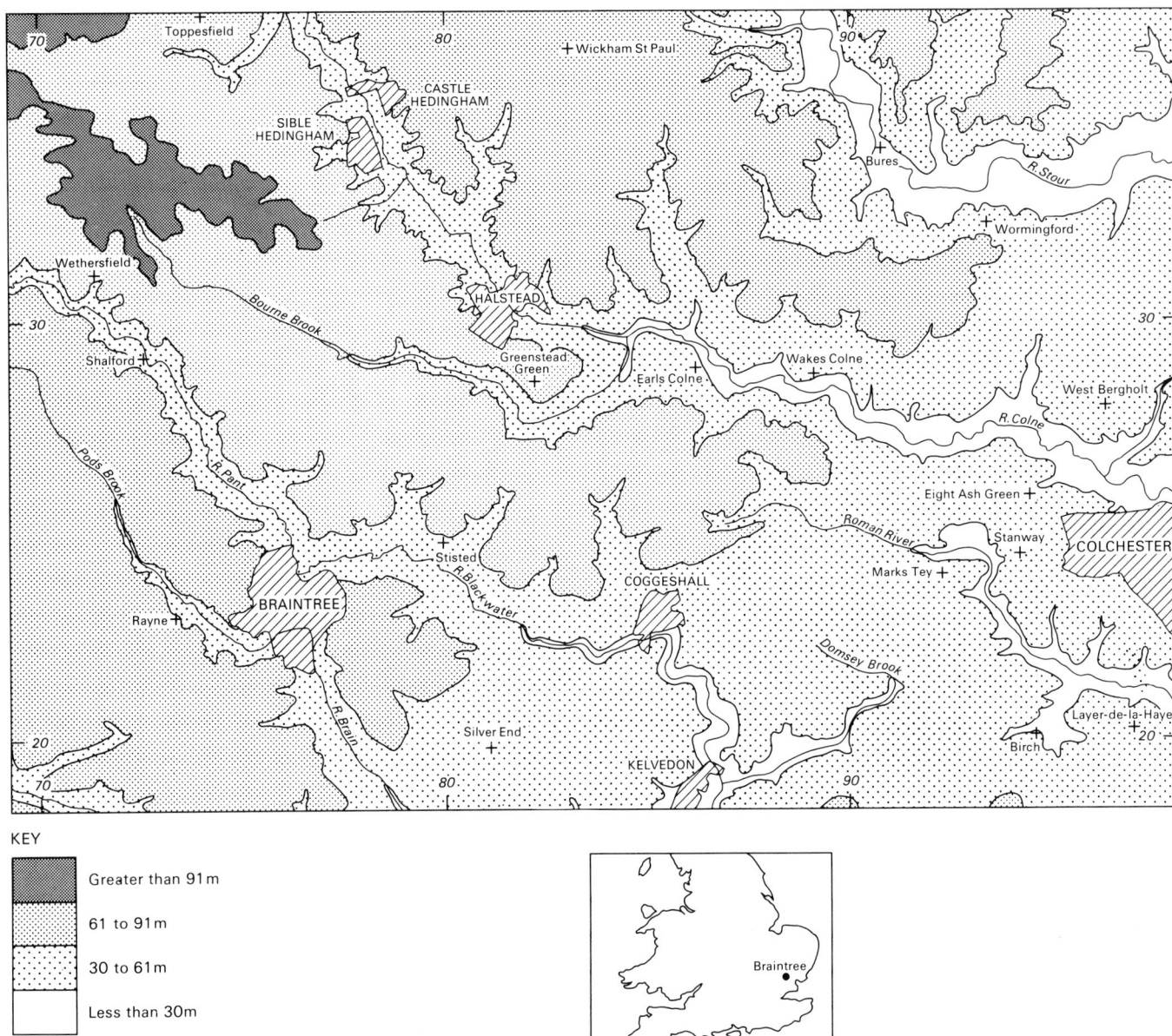

KEY

▓	Greater than 91m
░	61 to 91m
·	30 to 61m
	Less than 30m

Heights are above Ordnance Datum

Figure 1 Topography of the Braintree District

and Harvey, 1965). The hydrogeology of the district is also portrayed on the 1:250 000 Hydrogeological Map of East Anglia (Sheet 2).

Stamp (1921) made use of the early lithostratigraphical work on the Tertiary strata, and widened its scope to include a comparison with sediments of the same age in the Hampshire, Paris and Belgian basins. In the same paper he postulated a number of cycles of deposition to account for the variety of sediments. Wooldridge (1923, 1926) wrote in similar vein, linking the lithological variations of the Tertiary beds in the London Basin to structural controls. Both authors make little direct reference to this district; there is more detail on districts immediately south and south-west covered by the Chelmsford (241) and Epping (240) sheets. Wrigley (1924, 1940) was, contemporaneously, undertaking an exhaustive study of the faunas of the Palaeogene of the

London Basin. Subsequently a regional synthesis of the Lower London Tertiaries was published by Hester (1965), and a similar review of the London Clay by Davis and Elliot (1957). Curry (1958), on the basis of all previous research, formalised the Palaeogene nomenclature, and later wrote a general account (Curry, 1965) of the Palaeocene and Eocene strata in the London and Hampshire Basin.

Investigations into the early Quaternary deposits, which include the Crag, were carried out by Wood (1867) and Prestwich (1890). There is a short account of the Crag in the memoir for Sheet 47 referred to above (Whitaker and others, 1878). A little later, Harmer (1900) produced his classic account of Crag sedimentation, based largely on faunal evidence. His conclusions have not been radically altered to the present day.

Research into the later glacial and post-glacial deposits

followed much the same lines as the work on the early Quaternary. Wood (1867) recognised the presence of what he called 'subaqueous' glacial deposits, and subdivided them into Boulder Clay and Glacial Sand and Gravel. These divisions were mapped by the Geological Survey, and reported upon by Whitaker and others (1878) and Dalton (1880) in Geological Survey memoirs. Prestwich (1890) and later Solomon (1932) took a different view, considering that the glacial sand and gravel of earlier workers was a fluviatile or marine unit, which they called the Westleton Beds. Since then the Quaternary deposits of the Braintree district and adjacent areas have received little attention, but further subdivisions of the Quaternary sequence have been made on the basis of exposures on the coast of Suffolk and Norfolk. More recently, Clayton (1957) referred once again to a glacial origin for the sand and gravel, while West (1963) summarised the existing state of knowledge and formally drew up a stratigraphical sequence based primarily on the palaeobotanical investigation of interglacial deposits which he used to subdivide the sequence. A major contribution to this theme was produced by Turner (1970) who described Hoxnian interglacial deposits from Marks Tey in this district. A formal nomenclature relating to the entire Quaternary was subsequently produced for the Geological Society (Mitchell and others, 1973), and this terminology is used in the present memoir.

GEOLOGICAL HISTORY

The Braintree district is underlain at depth by the London Platform, part of a basement block of Palaeozoic rocks which has been relatively stable since Carboniferous time. Beneath this district the rocks which comprise the basement are probably of Devonian or Silurian age. Regional evidence from boreholes in London and southern East Anglia suggests that pre-Lower Cretaceous rocks are absent, having been removed from the London Platform during successive erosive periods. If any such rocks are present, they are likely to occur preserved in grabens let down into the basement rocks, a process postulated by Owen (1971) to explain the occurrence of Jurassic sediments in the area of the Thames Estuary.

The earliest Mesozoic sediments known to exist in this district formed during the mid-Cretaceous transgression, when Gault was laid down across the completely submerged London Platform. The advance of the sea continued with deposition of the Chalk in water generally some 200 to 400 m deep, though with short episodes of shallowing and associated gentle current action, during which, however, there was an extremely restricted influx of terrigenous sediment. Chalk deposition ceased with a period of uplift and emergence associated with mild deformation, when as much as 200 to 300 m of Chalk may have been eroded away from the Braintree district.

After some 15 to 20 million years the sea re-advanced across a slowly subsiding basin centred approximately along the Thames Estuary. The Tertiary sediments are clastic rocks laid down in this basin whose rate of subsidence and filling was partly governed by the larger and more rapidly deepening basin centred in the North Sea. The oldest Tertiary strata, the Thanet Beds, are bioturbated sands laid down in a shallow sea. Following a hiatus in sedimentation, the Woolwich facies of the Woolwich and Reading Beds overlapped the Thanet Beds and in some areas were channelled into them. The subsequent Reading facies of the Woolwich and Reading Beds comprise clays and sands deposited in a fresh or brackish-water lagoon or marsh which extended across Essex and North London to Berkshire and Hampshire. The distinctive colour-mottled appearance of these strata bears witness to the emergent conditions and to a contemporaneous fluctuating water-table. A further marine transgression followed, with the deposition of sandy and pebbly beds of the basal London Clay over most of the district. There was then a progressive deepening of the basin during which the argillaceous London Clay accumulated in anaerobic waters. The rather cold-water benthonic fauna of the London Clay, suggestive of sea depths of up to 350 m (Curry, 1965), contrasts with its flora which is derived from a tropical rain forest thought to border the sea during this period (Chandler, 1961). Widespread volcanicity occurred early in London Clay (lower Eocene) times in the north-west of the British Isles and near Denmark, while contemporaneous ash showers fell over the Braintree district (Knox and Ellison, 1979). The upper part of the Eocene succession, represented by the Claygate Beds and Bagshot Beds to the south (Bristow, 1985), is absent in the Braintree district. The deposits were removed by erosion triggered by uplift associated with earth movements, which were at a maximum in the Miocene and which caused minor flexuring and faulting. There is no further evidence of sedimentation in the Braintree district until earliest Pleistocene time, about 1.2 million years ago, when a marine transgression laid down the shelly sands of the Crag on the eroded upper surface of the London Clay. A considerable thickness of London Clay, possibly up to 50 m, had been removed in the intervening time, and the magnitude of the overstep is illustrated to the north of the district where the Crag transgresses beyond the London Clay on to Palaeocene strata and finally on to the Chalk. The Crag fauna contains a high proportion of cold or boreal species thought to indicate the oncoming of a climatic deterioration that heralded the climatic fluctuations from warm temperate to glacial that have lasted until the present day.

Most of the Pleistocene strata in this district are 'drift' deposits. The earliest are the Kesgrave Sands and Gravels, which were laid down in a braided river system prior to the arrival of the Anglian ice-sheet some 500 000 years ago. As the ice-sheet approached, poorly sorted sand and gravel deposits, the Glacial Sand and Gravel, associated with meltout from the ice-sheet, were laid down as torrent gravels and mudflows from heavily laden braided rivers. They were in part overridden by the ice front. When the ice sheet finally ablated, a chalky clay lodgement till, (described as Boulder Clay in this memoir), was left as a blanket over almost the entire district. Much of the meltwater was confined into narrow valleys, some of which were deeply channelled and appear to be ungraded; they are now commonly filled with Glacial Sand and Gravel outwash as well as till. The position of these meltwater channels commonly coincides with the major valleys of the present drainage system. Two of the over-deepened outwash routes, at Marks Tey and Kelvedon, became the sites of lakes when the ice melted. At Marks Tey, varved lake sediments contain pollen which demonstrates a transition from a cold period (the Anglian), through a succeeding temperate interglacial period (the Hoxnian), and back into cold conditions (the Wolstonian). By estimating the number of annual varves, Turner (1970) has calculated that the duration of the Hoxnian Stage at this locality was about 25 000 years. Later glacial phases are not represented by tills in this district, and are substantiated only by cryoturbation structures and solifluxion deposits.

In late Anglian times, till and outwash deposits filled some of the pre-existing valleys and caused modifications to the pre-glacial drainage pattern. River erosion continued in the valleys; most of the till, sand and gravel has now been eroded away and only patches of gravel remain, forming flights of terraces in each valley. The lowest of these are probably Ipswichian and early Devensian in age. In the latest Devensian, and throughout the Flandrian, during a period of generally rising sea-levels, the river valleys have been partially filled with argillaceous sediments which are still accumulating as alluvium, forming flood plains along all the rivers in the district which range in width from less than 100 m in Pods Brook to almost a kilometre in the Stour valley.   RAE

STRUCTURE

The district lies on the northern limb of the broad syncline known as the London Basin which closes westwards in the vicinity of Newbury and links with the southern North Sea Basin to the east. Although Wooldridge (1923) has described minor structures in adjacent areas, no significant folds are apparent from the current appraisal of borehole records. The Chalk surface, which approximates to a stratigraphical datum-plane, falls gently south-south-eastwards from 53 m OD near Howe Street [695 343] to −60 m OD in the southeast. A small fault at St Botolph's Bridge [9736 2730] is associated with a minor indentation of the structure contours on this surface. The presence of this fault was confirmed by a BGS borehole [9716 2734], which proved London Clay to 10.8 m above OD, whereas site investigation boreholes close to the south-east [9739 2725; 9752 2723] found that the top of the Upper Chalk here lies at levels of 1 to 2 m OD (see Figure 4). The trend of the fault cannot be determined from the present borehole evidence, but it may well approximate to that of the valley. Movement along this fault is unlikely to have been responsible for the Colchester earthquake of 1884 (Davison, 1924, pp.338–341), which apparently had an epicentre to the east and south-east of Colchester but the fault responsible for the earthquake was estimated to trend at 28° (Anderson, 1942, p.135), much the same as that suspected for the fault at St Botolph's Bridge.   RDL

CHAPTER 2

Concealed formations

PALAEOZOIC

There are no deep borings in the Braintree district and the Palaeozoic strata have not been proved. Nevertheless, they are thought to be present at a relatively shallow depth, and an estimation of their age and depth of burial can be attempted by extrapolation from deep borehole records in London and adjacent areas in southern East Anglia (Figure 2). Such boreholes have proved rocks ranging from Carboniferous to Lower Silurian (Llandovery) in age. The Palaeozoic rocks are part of a stable block, the London Platform, whose peneplaned surface beneath the Mesozoic cover falls from around – 200 m OD in the north-west to below – 300 m OD in the south. To the west of this district, the Ware Borehole proved 10.7 m of mudstone, limestone and calcareous sandstone assigned to the Wenlock; Devonian rocks were encountered at Turnford, Little Chishill and Ashwell [TL 286 390]. To the east, boreholes at Weeley and Stutton penetrated cleaved grey shale (Llandovery) and shaly sandstone (Upper Llandovery or Wenlock), respectively (Lister, 1971). There is no evidence to suggest that rocks other than of Devonian or Silurian age form the London Platform beneath the Braintree district.

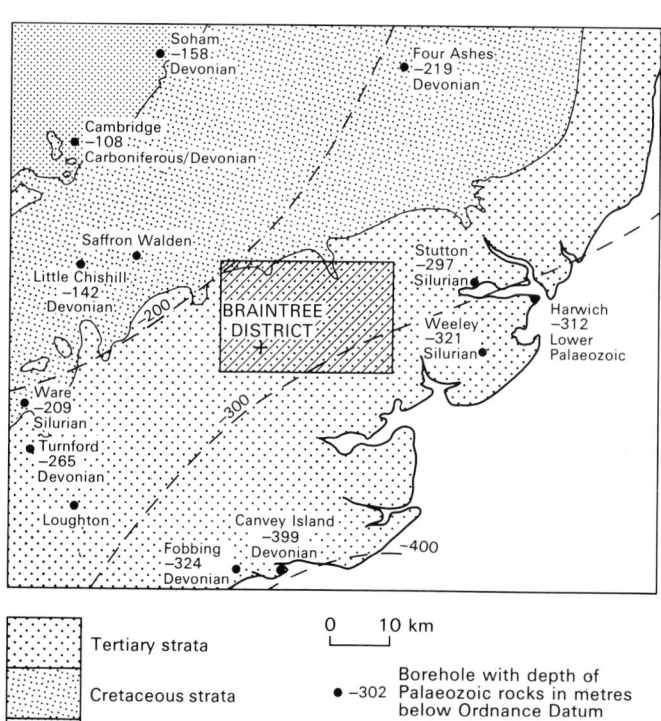

Figure 2 Contours, in metres below Ordnance Datum, on the surface of the Palaeozoic rocks in the Braintree and adjacent districts

MESOZOIC

Jurassic

The feather edge of the Jurassic strata overstepping on to the northern part of the London Platform has been proved in boreholes to the north-west at Ashwell and possibly at Saffron Walden. No Jurassic strata have been proved immediately adjacent to the Braintree district, though some may be preserved in grabens.

Cretaceous

LOWER GREENSAND

The extent of these strata is probably similar to that of the Jurassic, a thin sequence being proved at Ashwell and questionably at Saffron Walden. A discussion of the latter occurrence appears in the memoir relating to that district (White, 1932). In the Braintree district the only likelihood of preservation of the Lower Greensand, as with the Jurassic, would be in a downfaulted block or similar structure.

GAULT AND UPPER GREENSAND

Gault has been proved in all the deep borings in the area adjacent to the Braintree district. It is, therefore, almost certainly present at depth in the district and may be some 30 to 50 m thick. The maximum thickness of Gault proved in the surrounding borings was 50.8 m at Ware where it was glauconitic at the base; by comparison, 46.8 m of grey silty clay with phosphatic nodules were penetrated at Cheshunt. In the Saffron Walden district up to 47 m of Gault are present, consisting of 'blue, lead-coloured or pale grey tenaceous clays and marls, partly homogeneous, partly laminated and containing many phosphatic nodules reflecting periodic reworking of the sediments, and pyrite nodules' (White, 1932). Thin bands of hard shelly limestone also occur, and there is a glauconitic loamy sand with phosphatic nodules up to 3 m thick at the base. South of the Braintree district, 35.9 and 34.0 m of Gault were proved at Fobbing and Canvey respectively, but a rather thinner sequence is present to the east where 23.2 m were proved at Weeley and 18.6 m at Harwich. A combined thickness of Gault and Upper Greensand of only 15.3 m was penetrated at Stutton.

Where the Upper Greensand can be separated from the Gault it is thin, with a maximum known thickness of 13.4 m at Cheshunt. At Weeley it is absent, and is very thin or absent at Harwich. To the north-west, similar but younger sands are termed the Cambridge Greensand. A thin condensed sequence, it consists of a speckled sandy marl containing phosphatic nodules and phosphatised fossils, passing up conformably into the Chalk. A more detailed description of these strata is given in the memoir for the Saffron Walden district (White, 1932, p.12), where they crop out. Extrapolating from the known deposits, any sand present at this level in the Braintree district is likely to be less than 5 m in thickness, and of a similar lithology to the Cambridge Greensand.

CHALK

The Lower, Middle and Upper Chalk are the main divisions generally recognised in south-eastern England. The subdivisions, stages and zones of the Chalk likely to be present in this district are given in Table 1.

Table 1 Subdivisions of the Chalk

Stratigraphical division	Zone	Stage
Upper Chalk	Marsupites testudinarius	Senonian
	Uintacrinus socialis	
	Micraster coranguinum	
	Micraster cortestudinarium	
Middle Chalk	Sternotaxis planus	Turonian
	Terebratulina lata	
	Mytiloides labiatus	
Lower Chalk	Metoicoceras geslinianum	Cenomanian
	Calycoceras guerangeri	
	Acanthoceras jukesbrownei	
	Acanthoceras rhotomagense	
	Mantelliceras dixoni	
	Mantelliceras mantelli	

The Chalk does not come to crop in the Braintree district, but Upper Chalk lies directly beneath Quaternary deposits in the north-west; its subcrop extends down the Colne valley as far as Sible Hedingham and down the Stour valley to Bures. In the eastern part of the district, the Upper Chalk is overlain by Palaeogene deposits. The higher part of the Upper Chalk succession has been removed by denudation prior to the deposition of the Palaeogene sediments and by post-Eocene erosion, the youngest zone proved being the *Marsupites testudinarius* Zone in the north-east of the district. Lithologies typical of the Lower, Middle and Upper Chalk are described in surrounding districts (see, for example, White, 1932 and Worssam and Taylor, 1969). The Lower Chalk is off-white, greyish or yellowish, and contains some marl and clay but no flint. The Middle Chalk is off-white and somewhat marly, with numerous discrete marl partings; the base of the succession is nodular and pebbly, and nodular chalk reappears towards the top, together with scattered bands of nodular flint. The Upper Chalk is almost pure white with some bands of nodular chalk and mineralised hardgrounds near the base, and is characterised by flints at most levels.

In the Braintree district, Chalk has been proved in many boreholes, but none of these has reached its base. An estimate of the thickness of Lower and Middle Chalk can be made by comparing their thicknesses in boreholes in adjacent districts. The Lower and Middle Chalk are relatively uniform in thickness and 50 and 66 m respectively are likely to be present. Progressively younger zones appear in a north-eastward direction. A (minimum) thickness of the preserved Upper Chalk can be established in the deepest well, which lies in the western part of the district at Shalford [7343 2808]. Here 115 m of Chalk were proved of which

108.34 m are assigned to the Upper Chalk. At the base of the Upper Chalk is 0.45 m of Chalk Rock (given erroneously as the Melborne (*sic*) Rock (Sayer and Harvey, 1965, p. 106)). Thus, taking into account the estimated Middle and Lower Chalk figures, the total thickness of the Chalk at Shalford is likely to be approximately 225 m.

Faunas from boreholes at Witham and Kelvedon, just beyond the southern limit of the district, prove that the Chalk underlying the Palaeogene sediments belongs to the highest beds of the *Micraster coranguinum* Zone or, in the case of Kelvedon, possibly to the overlying *Uintacrinus socialis* Zone (Wood *in* Bristow, 1985). Boswell (1929) recorded the *socialis* Zone at Ballingdon, some 3 km within the Sudbury (206) district. The *Marsupites testudinarius* Zone also occurs in the Sudbury district, where Boswell described it as 18 to 20 m of soft marly chalk with few flints; the overlying *Offaster pilula* Zone chalk was said to be harder with spindle-shaped and cylindrical flints.

In this district Mr C. J. Wood of BGS Biostratigraphy Group reports that biostratigraphical data for the sub-Palaeogene Chalk are provided by boreholes at Bowdens Farm, Wormingford [9406 3339] and at Popsbridge near Nayland [9706 3389], both in the north-eastern area.

In Bowdens Farm Borehole, 37.8 m of Chalk were sampled beneath Quaternary and Palaeogene deposits. *Marsupites testudinarius* calyx plates and secundibrachials were found at 23.2 m depth, and plates and secundibrachials of *Uintacrinus socialis* at 24.4 m, indicating the *testudinarius* and *socialis* zones respectively. The *socialis* Zone could be proved to 28.9 m and possibly to 30.5 m on the presence of the distinctive secundibrachials and, at the higher horizon only, the belemnite *Actinocamax verus*. The underlying Chalk, first sampled at 45.7 m provided *Conulus* fragments and columnals of *Bourgueticrinus granulosus* suggestive of a high level in the *Micraster coranguinum* Zone. In Popsbridge Borehole, 104.5 m of Chalk were drilled beneath Quaternary and 13.1 m of Palaeogene deposits. Elongate variants of *Pseudoperna boucheroni* at 20.4 and 23.2 m depth, together with a probable small corroded *Marsupites* plate at 21.9 m, suggest that the *testudinarius* Zone is present at the top of the preserved Chalk succession. The underlying *socialis* Zone is positively identified by a *Uintacrinus* radial plate at 35.05 m, but probably extends from 24 to 39.62 m to judge from the range of secundibrachials attributable to this genus. The occurrence of *Retispinopora lancingensis* at 39.62 m is of interest, since elsewhere this species characterises the top of the *Offaster pilula* Zone. *Actinocamax verus* occurs at the same depth, and also at 45.7 m where pieces of hard green-stained chalk and associated pyrite suggest the presence of a hardground or glauconitised surface. Fossils from between 45.7 to 48.8 m depth include *Parasmilia sp.*, *Bourgueticrinus granulosus*, *B. ellipticus*, abundant fragments of *Conulus sp.* and large coarsely spinose radioles of *Stereocidaris sceptrifera* and are indicative of the higher part of the *Micraster coranguinum* Zone. It is possible that the surface or hardground at 45.7 m could represent the Barrois' Sponge Bed close to the top of the *coranguinum* Zone in Thanet. The identification of the *testudinarius* Zone as the highest preserved zone at subcrop in both boreholes is in accord with the data from the adjacent Sudbury and Chelmsford districts, and with the sub-Palaeogene zonal map produced by Curry (1965, fig. 2). RAE

CHAPTER 3

Tertiary

The Tertiary sequence in this district lies unconformably upon the Chalk and beneath a widespread cover of Quaternary deposits. The beds were laid down in a slowly subsiding sedimentary basin centred in the region of the Thames Estuary but connected with the Anglo-Dutch North Sea basin (Rhys, 1974, fig. 8). The precise position of the basin margin during Tertiary times is uncertain, but it lay to the west of the Braintree district throughout the period; post-Eocene erosion has led to the removal of the Tertiary strata at the extreme north-western limit of the district. The deposits fall within the Palaeocene and Eocene Series. The placement of the limits of these series in Europe has been the subject of discussion in recent years, and there are differences of opinion as to where they should be in the London Basin. The main barrier to agreement is the fact that the English and north-western European stratotypes have not been satisfactorily correlated by either biostratigraphical or geochronological techniques. The chronostratigraphic stages involved are, in ascending order, the Thanetian, Sparnacian and Ypresian. There is general agreement that the base of the Eocene should be placed at or near the base of the Ypresian (Berggren, 1971; Fitch and others, 1978; Curry and others, 1978) although some dissenting authorities (Odin, Curry and Hunziker, 1978) take it at the base of the Sparnacian. The chronology of the English Tertiary has recently been discussed by comparing K-Ar radiometric dates obtained from glauconite in the sediments with a time-scale erected from the dating of Tertiary igneous rocks in the North Atlantic region. The age of the base of the Ypresian is estimated at 50 Ma by Odin and others (1978) and at 53 Ma by Fitch and others (1978). The same groups of workers also record differing figures for the base of the Sparnacian: these are 54 Ma and 56 Ma, respectively.

System	Series	Stage	Group	Formation
Tertiary	Eocene	Ypresian		London Clay
	Palaeocene	Sparnacian	Lower London Tertiary Group	Woolwich and Reading Beds
		Thanetian		Thanet Beds

Figure 3 Tertiary stratigraphical nomenclature, as used in this memoir

The relationship of these stages to the lithostratigraphical formations is also in some dispute. Three formations of the Tertiary are present within the Braintree district, namely, in ascending order, the Thanet Beds, Woolwich and Reading Beds, and the London Clay. In the centre of the Southern North Sea Basin, the Palaeocene–Eocene junction is taken at the base of a volcanic 'ash marker' (Curry and others, 1978), which has recently been discovered onshore at the base of the London Clay near Harwich and also at Bulmer, only 1 km north of the Braintree district (Knox and Ellison, 1979). In addition wider biostratigraphical correlation has been carried out using dinoflagellate assemblages (Knox and Harland, 1979). This evidence suggests that the Palaeocene/Eocene boundary should be drawn at the base of the London Clay as mapped in the Braintree district. Figure 3 attempts to show the equivalence of series, stages and formations within the Tertiary.

Mineralogical and petrological studies of the Tertiary deposits, using a combination of X-ray diffraction (XRD) techniques and optical petrography, have been carried out in association with the present survey. A total of 23 core samples were examined from two boreholes at Bures [9120 3399] and at Wormingford [9267 3262] drilled during the survey; detailed logs of both boreholes appear in Appendix 1. An outline of the mineralogical results has been included in the following account.

LOWER LONDON TERTIARY GROUP (PALAEOCENE)

Nomenclature and general stratigraphy

The Palaeocene rocks of the London Basin have been grouped together as the 'Lower London Tertiaries' since the early work of Prestwich (1852) who subdivided the sequence into 'Thanet Sands', 'Woolwich and Reading Series' and the 'Basement Bed of the London Clay'. These terms later became known as Thanet Beds, Woolwich and Reading Beds, and Oldhaven Beds, respectively (Whitaker, 1866). The nomenclature was then formalised by Cooper (1976), who included the Thanet, Woolwich and Reading, and Oldhaven formations in the Lower London Tertiary Group. Subsequently, although Curry and others (1978) considered that only the lower two of these formations fell within the Palaeocene, work on dinoflagellate assemblages by Knox, Harland and King (1983) indicated that the Palaeocene–Eocene boundary should be placed at the top of the Oldhaven Beds of north-east Kent. However, in the Braintree district this controversy is less relevant, as the Oldhaven Beds are not recognisable, although strata equivalent in age to them probably occur at the base of the London Clay. Indeed the poor quality of many of the borehole logs precludes any subdivision of the Lower London Tertiary Group on the 1:50 000 sheet, although the major subdivisions can be recognised in about half of the boreholes. Detailed information has fortunately been obtained from the BGS Bures and Wormingford boreholes.

The Lower London Tertiary Group is exposed at only two localities, namely at Sible Hedingham [785 340] and at Lamarsh [894 356]. The group extends at depth throughout most of the district, and the area of subcrop, as determined from borehole data, is shown in Figure 4. It stretches from

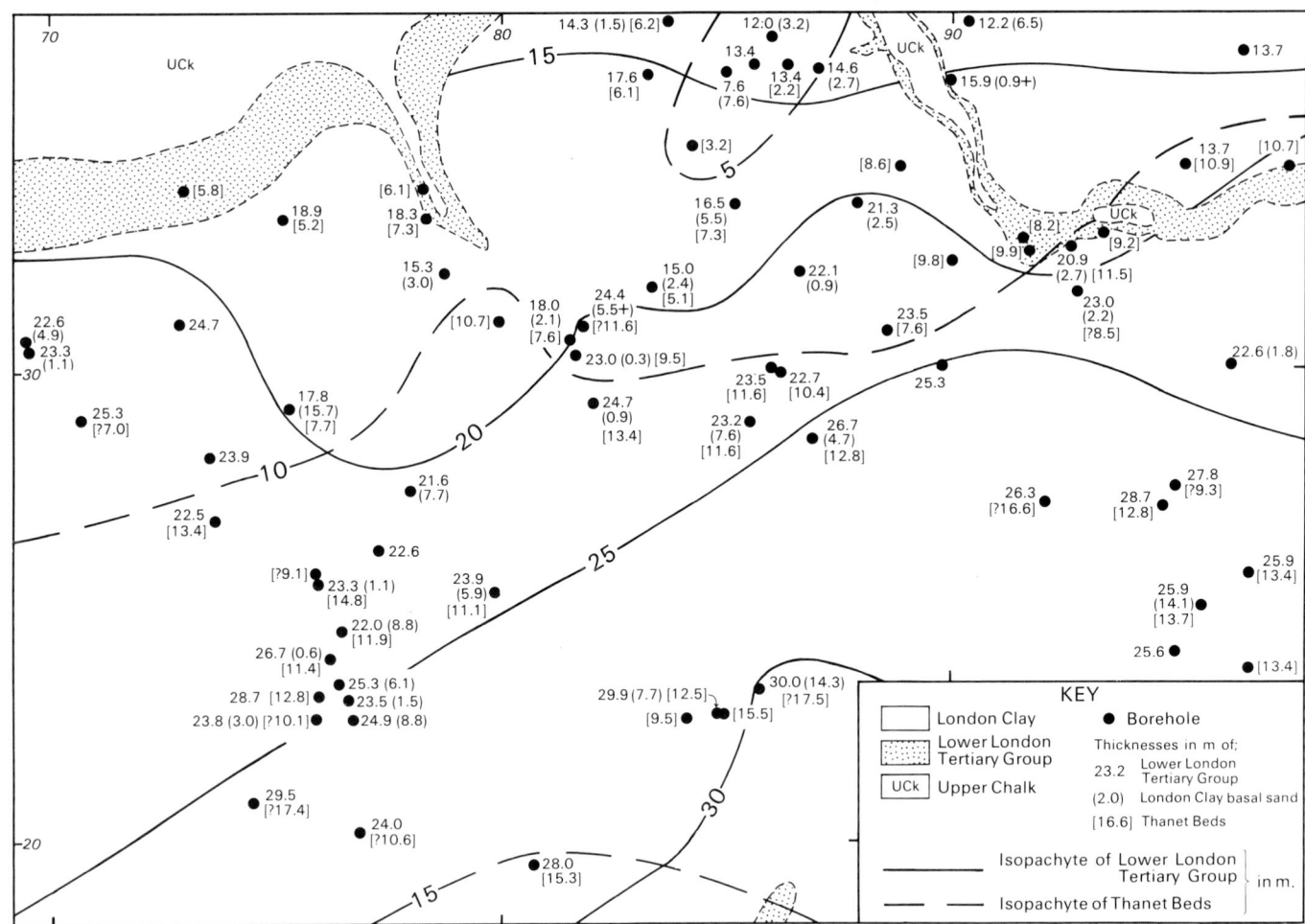

Figure 4 Isopachytes of the Thanet Beds and of the Lower London Tertiary Group

around Wethersfield Airfield in the west to Rushley Green [786 306], forms much of the floor of the Colne valley to the south of Hull's Mill Farm [792 332] and also underlies the Stour valley.

The base of the deposits falls gently in a south-easterly direction from around 53 m above OD to below 50 m below OD in the south of the district on a gradient of about 1 in 200 that shows only minor irregularities. The thickness increases in the same direction from less than 15 m in the north to a maximum of 30 m around Coggeshall [805 203] and is generally more than 25 m in the south-eastern part of the district (Figure 4).

THANET BEDS

Stratigraphy

The Thanet Beds are not exposed at the surface in the Braintree district, but they have been shown to crop out beneath the drift in the north-west and, in the south, along the Blackwater valley near Kelvedon [863 187]. They thicken southwards, from less than 5 m between Wickham St Paul [832 303] and Alphamstone [877 293], to more than 15 m at Fordstreet [920 270] and south of Silver End [800 197]. This thickening continues southwards across the adjacent Chelmsford district (Bristow, 1985, fig. 7). The isopachytes

(Figure 4) are necessarily generalised because of the wide range of thickness values obtained from borehole logs of variable quality; thus an apparently substantial variation in thickness over a short distance may arise from the apparent unreliability of a particular log.

The basal junction of the Thanet Beds with the Chalk is sharp and is always recognisable in borehole logs. The basal bed, known as the Bullhead Bed, generally contains nodular flints set in a green glauconitic sandy clay matrix. It can be recognised in many borehole records across the district, and is locally up to 2 m or so in thickness. In the Wormingford Borehole it consisted of 1 m of green and dark grey mottled and bioturbated silty fine-grained sand, from which diatoms of the genus *Coscinodiscus*, comparable with those recorded from the Thanet Beds in the Southend district (Lake and others, 1986) were recovered.

Above the distinctively coloured Bullhead Bed, the Thanet Beds are fairly homogeneous, consisting predominantly of bioturbated grey to greenish grey fine-grained silty sands. They are variously described in borehole logs as 'clays' 'loams' and 'sands', usually 'grey' in colour, but described as 'greenish grey', 'blue' or just 'dark' in some records. In more detail, the 11.5 m of Thanet Beds proved in the Wormingford Borehole consisted of well sorted, silty fine-grained sand with angular to subangular quartz grains, slightly glauconitic in places, particularly at the base, and varying in colour from grey (N4) to brown (5 YR 3/1). The top part of the formation was laminated, but bioturbation was observed in the lowest 5 m. Thin intercalations of green and reddish green mottled clays occurred in the top 2 m. Only the uppermost 7.2 m of the Thanet Beds were recovered from the Bures Borehole. They consisted of dark greenish grey bioturbated micaceous fine-grained sand with scattered glauconite pellets; no fauna was recorded.   RAE

Mineralogy

A significant feature of the Thanet Beds is the widespread presence of volcanogenic materials. Fresh pyroclasts of basic to intermediate lava occur in the basal glauconitic beds and much of the overlying sand is zeolite-bearing. Well-crystallised smectite is abundant in clay fractions, and its close association with zeolite suggests it also is volcanogenic. Zeolite distribution suggests that at least 60 per cent of the total thickness of Thanet Beds contains some volcanic ash, altered *in situ*, the concentration being greatest near the base. The zeolite is a member of the clinoptilolite-heulandite series (hereafter termed clinoptilolite for brevity), and is found throughout the lower and middle parts of the Thanet Beds in the Wormingford and Bures boreholes (see Figure 5), and also in the Great Cornard (Ellison, 1976) and Great Bardfield (Ellison and others, 1977) boreholes in adjacent districts. A similar mineral has been described from the Thanet Beds in Kent, where it is associated with pyroclastic material (Brown and others, 1969; Knox, 1979). This same association is found in the basal bed in Wormingford Borehole (32 to 33 m depth): a thin section (E 51375) shows grains of lava, glauconite, feldspar and flint set in a brown-stained, weakly birefringent matrix which forms about 50 per cent of the sediment. Rounded lava particles, ranging from 0.05 to 0.2 mm in size, form about 10 per cent of the thin section. They show glassy and micro-crystalline textures, the latter consisting of tiny plagioclase laths, in places crudely aligned, embedded in a colourless or pale green groundmass. Extinction angle measurements are possible on a few large plagioclase fragments and indicate an andesine-labradorite composition; similar results were obtained by XRD techniques. Such a composition suggests an intermediate or basic parent lava. The composition of the glass is indeterminate, for it is invariably devitrified and largely altered to glauconite. Clinoptilolite mostly occurs as colourless aggregates of silt-sized anhedral grains intergrown with the smectite-rich matrix. Intergrowths of tiny prismatic clinoptilolite crystals also occur as a cement in some clay-free areas, possibly representing infilled burrows, and also in the solution moulds of *Geodia* spicules and radiolarian tests. The zeolite is clearly of authigenic origin, and its association with lava particles and plagioclase fragments suggests a genetic relationship, the zeolite forming as a result of alteration of fine glassy ash. Much of the clay matrix probably has a similar origin.

Clay minerals are a significant constituent of the Thanet Beds, both as clay matrix minerals and as glauconite grains. Throughout the Thanet Beds smectite dominates the clay matrix, commonly forming 60 to 70 per cent of the $<2\,\mu m$ fractions which in turn make up between 20 per cent and 50 per cent of the sediments. Illite forms up to 30 per cent of the $<2\,\mu m$ fractions, but is much less abundant in the basal bed. Chlorite is present as a minor constituent of the clay fractions, but kaolinite is rarely present, even in minor amounts. The smectite is characteristically well crystallised, using the criteria of Biscaye (1965), throughout the Thanet Beds and commonly occurs in silt-sized aggregates of clay flakes. These aggregates almost certainly result from argillisation of ash particles *in situ*, and their occurrence may account for the silty appearance of the Thanet Beds.

Glauconite grains are particularly abundant in the basal bed where they may form as much as 40 per cent of the sediments. Elsewhere in the Thanet Beds glauconite occurs as scattered grains amounting to a few per cent only. Three types of glauconite are recognised in the basal bed in the Wormingford Borehole. The most common (type 1) consists of subangular to subrounded grains showing a variety of colours from deep green to yellowish or brownish green. Typically they show heterogeneous textures and some contain ghosts of feldspar laths or relict vitric structures, indicating a replacement origin. The second type (type 2) is generally larger (up to 0.4 mm). Deep green lobate pellets show homogeneous textures and desiccation cracks filled with clay and zeolite. XRD data suggest that both types are interlayered glauconites (Class 2 of Bentor and Kastner, 1965). The subangular and subrounded grains with heterogeneous texture and the deep green lobate pellets are both composed of a dioctahedral mica/smectite mixed-layer mineral characterised by a very broad peak between 10 Å and 12 Å on X-ray powder patterns (X 7958, 7959). This peak is broader and more diffuse in the case of type 1 glauconites, suggesting that these grains contain more interlayered smectite than the lobate pellets. A third type occurs as pigmentary glauconite (Triplehorn, 1966). It is well known in hand specimen as the green coating to the typically nodular flints of the Bullhead Bed as well as on well-rounded flints. It has a composition similar to the type 1 glauconite and also forms a thin coating on lobate glauconite pellets (type 2), flint and lava particles, other clastic grains and, less commonly, on glauconite grains of type 1.

Above the dominantly glauconitic and tuffaceous basal bed, the Thanet Beds become progressively more orthoclastic and arenaceous, the dominant composition being clayey fine-sand, though for reasons noted above the sands have a silty appearance in hand specimen. Scattered flakes of muscovite and, less commonly, chlorite are present in the sands, the occurrence of these minerals coinciding with the increased abundance of illite and chlorite in the clay fractions. Orthoclastic feldspars, albite and microcline also become increasingly abundant at the expense of volcanogenic plagioclase. These mineralogical trends indicate a steady reduction in the amount of ash accumulating in the Thanet Beds, with by far the greatest concentration occurring in the basal beds.   RJM

WOOLWICH AND READING BEDS

Stratigraphy

These strata crop out at Sible Hedingham [785 340]. The area of subcrop cannot be accurately ascertained from the limited borehole information available. Nine wells in this district have proved Woolwich and Reading Beds immediately beneath drift deposits in the northern subcrop. A narrow subcrop beneath drift in the Blackwater valley in the south of the district is inferred from boreholes in the adjacent Chelmsford district (Bristow, 1985, fig. 8).

The Woolwich and Reading Beds as a whole consist of two main lithological types: green coloured glauconitic sands dominate the lower part of the sequence, whilst the upper part is generally a red mottled clay. These two lithologies respectively correspond broadly to the 'Bottom Bed facies' and 'Reading facies' of Hester (1965). They both form continuous units across the district but it is not possible to separate them in many of the borehole records, either because of the poor quality of the logs or the localised interdigitation of the two lithologies. Between about 8 and 15 m of sediments are present over most of the district and there is no detectable thinning towards their limit in the north-west.

The Woolwich and Reading Beds everywhere rest on Thanet Beds. Their base is taken at an upward change to dominantly green glauconitic sands (5GY to 10GY hues), which usually coincides with an increase in grain size and, locally, with the presence of a thin bed of rounded black flint pebbles. The basal bed in well logs is usually described as a bed of 'greensand' or, less commonly 'a green clay'. The Bures Borehole showed a gradational change from Thanet Beds to Woolwich and Reading Beds with a gradually increasing glauconite content and a coarsening of sand. The base was taken at a change from bioturbated glauconitic fine-grained sand (Thanet Beds) to finely laminated glauconitic fine- to medium-grained sand exhibiting alternate dark green and pale green laminae. There was a sharply defined base in the Wormingford Borehole: the basal bed consisted of green and red mottled glauconitic clayey sand, contrasting with the grey slightly bioturbated fine-grained sand of the underlying Thanet Beds.

The upper part of the Woolwich and Reading Beds consists dominantly of clays, with minor amounts of sand and silt, which in borehole logs are described variously as 'dark', 'blue', 'red', 'coloured' and 'mottled'. The most characteristic lithology is a stiff to hard clay of red to purple hues (10R to 5P), with vertically oriented colour veining and mottles of orange (5Y 4/2) and pale greyish blue (5B 5/1), possibly reflecting original bioturbation as well as root casts. Sub-vertical shears with listric surfaces commonly occur in these clays. Silty beds in this upper part of the formation commonly display fine lamination, and small burrows and bioturbation are recorded. In the more silty and fine sandy beds (e.g. 14.0 to 15.3 m in Wormingford Borehole), optical examination revealed moderately sorted subangular to subrounded quartz grains commonly exhibiting strains characteristic of a metamorphic rock, along with a few polycrystalline grains of quartzitic sandstone and rutilated quartz. Up to 10 per cent of the grains are of feldspar with minor amounts of flint fragments and glauconite pellets. Chlorite and mica intergrowths coat many of the feldspar and quartz grains, indicating derivation from low grade metamorphosed sediments.

A few indeterminate ostracods were recovered from Wormingford Borehole at 2.2 m below the top of the Woolwich and Reading Beds in a greyish brown fine sandy silt; no other fossils have been recorded from the formation within this district.   RAE

Mineralogy

There is an important mineralogical change between the clay assemblages in the two main lithologies. The lower beds contain clay assemblages which are richly smectitic and may be of volcanic origin, while smectite is significantly less abundant in the upper beds being replaced by illite as the dominant clay mineral: kaolinite, rarely present in the underlying Tertiary strata, is notably more plentiful in the upper beds (Figure 5).

The lower beds (4.4 and 4.5 m thick respectively) in the Bures and Wormingford boreholes (Figure 5) consist of bright green glauconitic medium-grained clayey fine- and medium-grained sands, with some mottling of olive-green (10G 5/2) and brown (10YR 5/6) due to alteration of glauconite to goethite and/or limonite; they may also show red or purple (10R 3/4) staining mostly due to hematite. Quartz sand and glauconite are the main constituents. A pale green clay matrix is present. Detrital quartz grains, ranging from 0.1 to 0.3 mm in size, are typically subangular and well sorted, and form 25 to 50 per cent of the sediment. The sand-grade material also contains flint fragments which contribute 5 to 10 per cent to XRD estimates of quartz content (Tables 4 and 5 in Appendix 2). These flint fragments generally show the same size range as quartz grains, though rare fragments reach 0.5 mm across; some are completely altered to glauconite, but retain the angular conchoidally fractured shapes typical of flint detritus. Minor amounts of feldspar (1 to 2 per cent) are generally present, and colourless flakes of muscovite and broken crystals of ilmenite are common accessory minerals. Glauconite grains form up to 30 per cent of these sediments and, together with the green clay matrix, are responsible for their striking colour. The morphology of the grains is similar to that seen in the Thanet Beds (see p. 9); most are angular and vary from pale to deep bottle green with a size range of 0.15 to 0.35 mm. Deep green, lobate grains are less abundant but generally largest, and commonly show desiccation cracks filled with whitish clay, which is mainly discrete smectite (X 7834). The sand-grade material is loosely bound together by a pale green clay matrix which may form up to 50 per cent of the sediment, and is composed of approximately equal amounts of smectite and illite which occur mostly as discrete minerals though some interlayering of the two minerals is detectable. The smectite is moderately well crystallised (V/p = 0.3) with a basal spacing of 15 Å which expands to 18.5 Å with glycerolation; it may be an Fe-montmorillonite or possibly nontronite. A broad, diffuse 10 Å reflection characterises the illite and resembles patterns given by associated glauconite grains.

In the upper part of this sequence the sandy beds with less glauconite (17.0 to 20.0 m depth in Wormingford Borehole and 19.0 to 21.0 m depth in Bures Borehole), but with a wide range of colour mottling, contain some clay beds consisting of up to 55 per cent of well crystallised dioctahedral smectite along with minor amounts of illite and kaolinite. No trace of zeolite or pyroclastic material has been found associated with

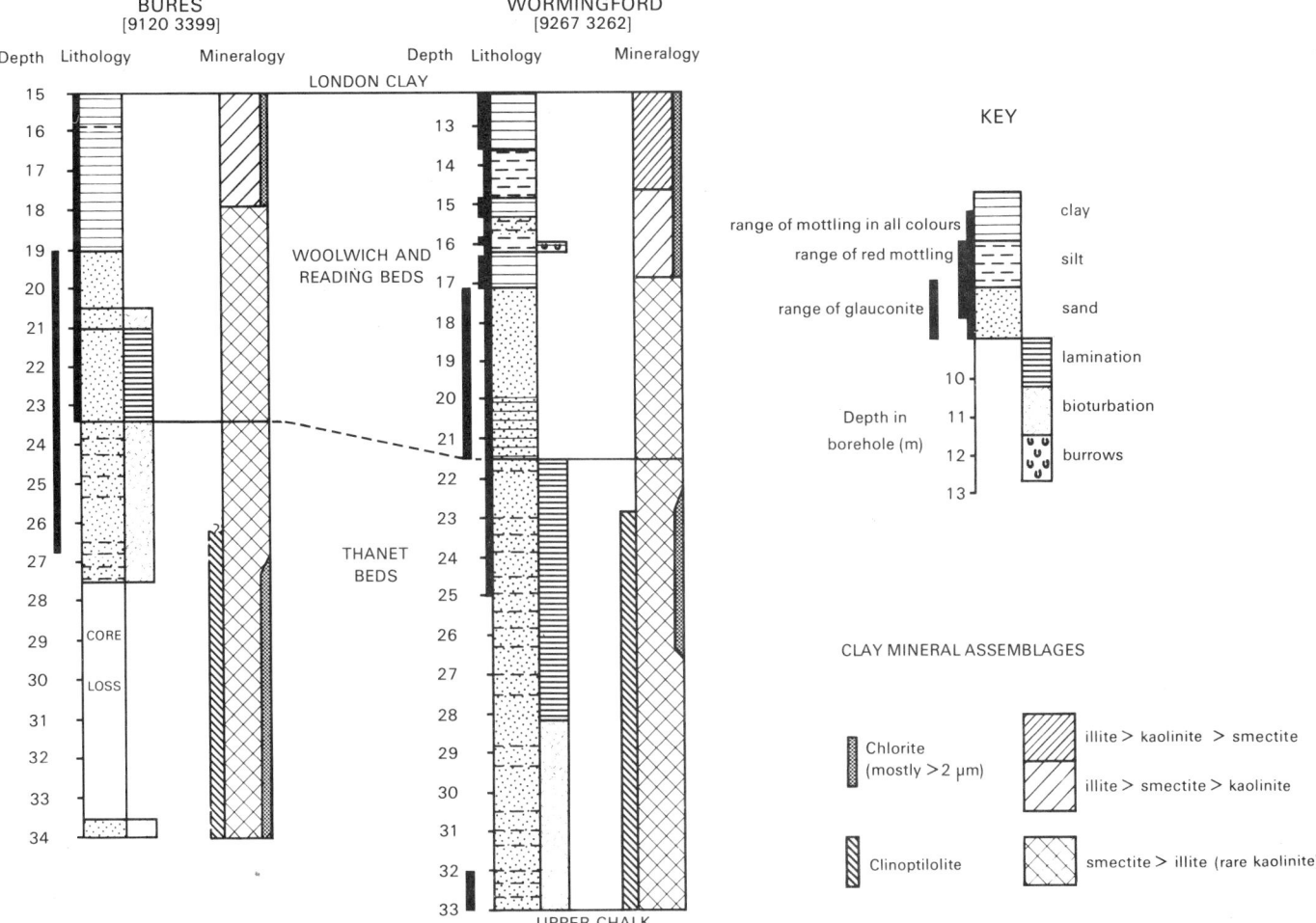

Figure 5   Lithological and mineralogical correlation of the Bures and Wormingford Boreholes

this smectite. In these beds, therefore, it seems less likely that the smectite has formed *in situ* from volcanic glass. On the other hand, because of the abundance and the well-crystallised nature of the smectite, derivation from a source of altered volcanic deposits is a distinct possibility. In the red mottled clays of the upper beds smectite is replaced by illite as the dominant clay mineral and kaolinite becomes much more abundant than it is in underlying Palaeocene sediments. Illite, smectite and kaolinite make up most of the <2 μm fractions, typical proportions being 50 to 65 per cent illite, 15 to 40 per cent smectite and 10 to 20 per cent kaolinite, with chlorite and a mixed-layer mineral occurring in minor amounts. Smectite crystallinity also changes in the mottled clay and is no longer the well crystallised variety found in the smectite-dominated assemblages of underlying beds.   RJM

London Clay (Eocene)

Stratigraphy

The Eocene strata of the eastern London Basin are divided into three formations; the London Clay, Claygate Beds and Bagshot Beds. The oldest formation, the London Clay, is the only formation occurring in this district. London Clay has

been studied since before the pioneering work of William Smith, who, in 1811, formally coined the term. The separation of the London Clay from clays which had previously been mistaken for it was made by Prestwich whose methodical work (1847; 1850) has formed the framework for all subsequent researches. An attempt to zone the London Clay on the basis of faunal changes in the sequence was made by Wrigley (1924; 1940) who summarised earlier efforts at subdivision. More recently, King (1970) has proposed a subdivision into six biofacies which were later modified by Stinton (1975). Further reference to the zonation of the London Clay in the Braintree district is made below. The lowest arenaceous beds of the London Clay were termed the Basement Bed of the London Clay (Prestwich, 1850) and subsequently re-interpreted by Whitaker (1866), who identified part of them as a separate formation, the Oldhaven Beds (Figure 3). In this district, similar sandy beds occur at the base of the London Clay; for convenience they are here grouped together and described with the London Clay.

London Clay is present beneath the whole district except in the extreme north-west, beneath the upper reaches of the Colne valley, in the Stour valley and in the lower part of the Blackwater valley (Figure 4). It crops out along the lower slopes of the major river valleys and their tributaries. One exception is in the Blackwater valley downstream from Coggeshall, where river downcutting has not breached the glacial deposits. A more extensive outcrop of London Clay occurs in the south-east of the district where it forms a low ridge between Messing-cum-Inworth and Layer-de-la-Haye.

Post-Eocene erosion has removed variable amounts of London Clay in this district. The London Clay has a gentle (less than 1°) southerly dip across the whole district and progressively higher beds are preserved towards the south. The maximum thickness proved was 62 m in a borehole at Black Notley [7683 2011].

A standard lithostratigraphic sequence for the whole London Clay sequence in Essex has been established recently in boreholes at Stock, 20 km south of this district (Bristow, 1985), and at Hadleigh in south-east Essex (Lake and others, 1986). Figure 6 shows four of the six lithological units (labelled LA to LF in ascending order) not to be confused with the biofacies recognised by King (1970; 1981) in these two boreholes, and compares them with the broad facies-controlled, biostratigraphical divisions which were proposed by Wrigley (1924) in the London area. The London Clay at outcrop over much of this district is probably referable to Unit LC of the above classification although there is no firm supporting lithostratigraphic or palaeontological evidence. Unit LD may, however, be present at two localities in the Braintree district: in the south-eastern corner, on the London Clay ridge at Layer-de-la-Haye [97 19], and also in the extreme south near White Notley [79 19].

The sandy beds, at the base of the London Clay in the north-west of the district are similar in lithology and mineralogy to units LA and LB of the standard sequence (Figure 6). Of particular similarity is the upward increase in smectite content at the expense of illite through the sandy and slightly micaceous clays of Units LA and LB; this increase levels off in Unit LC in silty clays which are relatively quartz-poor (less than 20 per cent) and show little variation in mineral proportions. Dolomite is characteristically absent from Unit LA and much of Unit LB in the Hadleigh (Sand Pit Hill) Borehole, and is also absent from similar strata in the Braintree district.

The London Clay rests disconformably on the Woolwich and Reading Beds. The junction crops out at only two localities in the district, at Sible Hedingham [787 340] and at Lamarsh [834 356], where the weathered basal London Clay consists of buff and pale brown silt and fine sand which contrast strongly with the stiff red mottled clays of the underlying strata. In the borings at Wormingford and Bures, the base was well defined in unweathered strata, the basal London Clay being a dark grey or greenish grey fine sand with rounded black flint pebbles, less than 1 cm diameter, overlying bright red mottled clays of the Woolwich and Reading Beds. In the Wormingford Borehole, the Woolwich and Reading Beds were burrowed to a depth of 2 cm and the burrows filled with medium- to coarse-grained sand. In well logs the junction is readily located; the basal London Clay beds are described variously as grey or brown 'sandy clay', 'loamy sand', 'dead sand' or 'running sand', and shelly sands and pebble beds are recorded in places. The thickness of these basal beds, which can be traced across much of the London Basin, is variable and exceptionally may reach 12 m.

In districts to the south, these sediments have been termed Oldhaven Beds, because of their resemblance to the type succession of Oldhaven Beds in Kent (Bristow, 1985; Lake and others, 1986). However, since their upper limit is not easily recognised in the present district, they are here included with the London Clay (see Figure 3).

A complete record of the lower part of the London Clay sequence was obtained from the boreholes at Bures and Wormingford; they penetrated 15.50 and 7.45 m of London Clay respectively. The lowest London Clay in these borings (see Appendix 1 for detailed borehole logs) consists of bioturbated dark olive grey (5Y 3/1) to greenish grey (5GY 3/1) micaceous silty fine-grained sand, which contains small rounded black flint pebbles up to 1 cm in diameter along with pyrite nodules, pyritised wood fragments, calcareous cemented siltstones with calcite veining, and buff iron-cemented siltstone. During the present survey, these lowest London Clay beds were not generally exposed, although buff to orange brown mottled clayey fine sands and silts, some 20 m above the base of the London Clay, were seen in a degraded brickworks section at Bures [901 341].

Progressing southwards across the district, successively higher beds of the London Clay crop out. In the south the typical lithology is an olive-brown to grey slightly micaceous silty clay which weathers to chocolate-brown on oxidation of the ubiquitous pyrite. Siltstones and septarian nodules were seen in small sections locally. Cementstones are present in the London Clay along the River Ter valley below Blake End [703 228] and its tributary near Bartholomew Green [720 210] to a point just north of Rutlands [711 197]. It is known from other areas (Lake and others, 1986) that these nodules usually occur at intervals along discrete bedding surfaces rather than randomly scattered throughout the sequence. Disturbance of the bedding above and below such nodules indicates they have formed during lithification, although Wrigley (1924) noted that in some areas the nodule surfaces exhibited contemporaneous boring. Small race concretions of re-precipitated calcareous material are common in the weathered clay. The material may be derived from the London Clay or, more likely in this district, from ground-

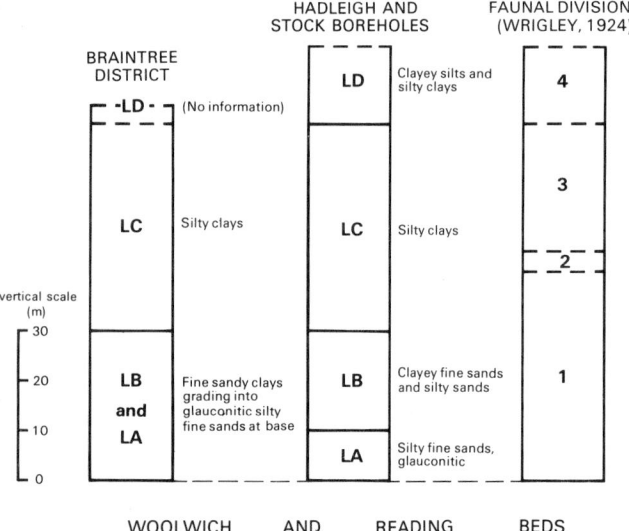

Figure 6   Correlation of the London Clay

water which has passed through overlying chalky drift deposits.

Selenite ($CaSO_4 2H_2O$), a product of chemical weathering by the interaction of calcareous material in the London Clay and sulphate ions in the groundwater from oxidation of pyrite, is found both in brown London Clay and also in the upper layers of grey London Clay. A white efflorescence of gypsum, precipitated by the weathering of finely disseminated pyrite, is common on borehole cores of London Clay only a matter of weeks old. The thickest oxidised weathering profile recorded during the present survey is 9.4 m in the Bures Borehole. Here London Clay crops out at the borehole site, and the sandy lithology has allowed relatively easy passage of groundwater. By comparison, less than 1 m of weathered London Clay is present beneath drift deposits greater than 5 m in thickness; numerous examples are detailed, for example, by Clarke and Ambrose (1975).

RAE

Mineralogy

The mineralogical composition of samples in the basal 3.5 m of London Clay in the Wormingford Borehole (Appendix 2, Table 4) shows that there is a general upward increase in the abundance of fine-sand size detritus, including quartz, feldspar, muscovite and chlorite, at the expense of clay minerals, but this trend is accompanied by a slight upward-fining of the grain-size of the sandy material. Samples of the same lithology from the Bures Borehole, taken 4 to 5 m above the base of the London Clay, show textures and compositions in accord with this trend. The total quartz content ranges from 20 to 25 per cent at the base of the London Clay to 40 per cent at 5 m above the base in the Bures Borehole (Appendix 2, Table 5). Carbonates are absent, but glauconite is sparingly present as scattered, commonly bright green grains. Passing upwards into the London Clay from the Woolwich and Reading Beds, there is a significant increase in the amount of smectite, a trend which continues through the lowest 5 m of London Clay studied in the two borings. In comparison kaolinite proportions are reduced, but show little variation within the London Clay, where typical clay fractions are illite 44 to 61 per cent, smectite 26 to 41 per cent, and kaolinite 13 to 15 per cent.

RJM

CHAPTER 4

Quaternary Solid: Crag

Solid and Drift strata have traditionally been separated on Geological Survey maps and in memoirs. The origins of the term 'drift' lie in the formative years of geological science. In general terms, the base of the drift deposits is now generally taken at the bottom of the oldest non-lithified deposits which are terrestrial in origin. In East Anglia the boundary between Solid and Drift, using the above definition, lies within the Quaternary sequence; in the present district the Crag is thus considered to be Solid, and is the oldest Quaternary deposit in the district.

A subdivision of the Crag was made by Harmer (1900) from an extensive study of its molluscan fauna in the many quarries and coast sections in North Essex and Suffolk. More recent work has suggested, however, that Harmer did not allow sufficiently for faunal variations resulting from facies changes in his Red Crag across its outcrop area, and it has been found that his subdivisions cannot be traced laterally. In this account deposits which may previously have been regarded as Red Crag are termed Crag. The sediments were arguably placed in the Pliocene (Boswell, 1929) but Mitchell and others (1973) formally placed the Red Crag at the base of the British Pleistocene (Table 2). Subsequently, however, Curry and others (1978) have indicated that part of it may be Pliocene.

Rusty brown shelly sands of the Crag were mapped during the early geological surveys in north-east Essex and Suffolk (Whitaker and others, 1878; Dalton, 1880), but none has been previously recognised in the Braintree district. In Sudbury, only 2 km north of this district, up to 4 m of Crag were well exposed in quarries during the survey by Boswell (Boswell, 1929).

During the present survey, the southern limits of the Crag has been extended into the Braintree district. The Crag is extensive north of the River Stour, with outliers west of the River Stour between Alphamstone [878 355] and Pebmarsh [854 335], and near Wethersfield [712 312]. Smaller isolated pockets occur near Great Maplestead [808 345], Wormingford [932 322] and Aldham [918 258]. Outcrops are limited to the valley sides though the Crag forms a persistent mappable unit only in the extreme north-east of the district, where it crops out at the bottom of a feature capped by the Kesgrave Sands and Gravels and Glacial Sand and Gravel. One exposure was noted during the survey, in a quarry [895 342] near The Ferriers at Bures where *Cardium? angustatum*, *Mya*, and barnacle valves have been recorded. The exact delineation of the Crag beneath drift deposits is conjectural but its general extent is clear from the outcrops and from borehole logs that record shelly sands. The base of the formation falls eastwards from over 50 m above sea-level to about 38 m. The maximum proved thickness is 4.7 m in the north-west.

The Crag is mostly composed of rusty to reddish brown, slightly silty, fine- to coarse-grained, subrounded sand, with minor amounts of gravel. Tabular iron-cemented beds,

Table 2 Subdivision of the Quaternary System, after Mitchell and others (1973).

System/Period			Stage	Formation(s) in this district
Drift		Holocene or Recent	Flandrian (Warm)	Alluvium, Peat
			Devensian (Cold)	River Terrace deposits
			Ipswichian (Warm)	River Terrace deposits
			Wolstonian (Cold)	
			Hoxnian (Warm)	Lacustrine deposits
	Quaternary	Pleistocene	Anglian (Cold)	Boulder Clay Glacial Sand and Gravel Glacial Silt
			Cromerian (Warm)	
			Beestonian (Cold)	Kesgrave Sands and Gravels*
Solid			Pastonian (Warm)	
			Baventian (Cold)	
			Antian (Warm)	(not represented)
			Thurnian (Cold)	
			Ludhamian (Warm)	
			Waltonian (Cold)	Crag

*Discussion regarding the Stage in which the Kesgrave Sands and Gravels should be placed is found on p.18.

generally less than 5 cm thick, are characteristic and are present in all lithologies from silts to gravelly sands. In four boreholes, the Crag has been encountered in its greenish grey unoxidised state. Shells are common in both the oxidised and unoxidised sediments. At outcrop the calcareous shell material has been leached away leaving casts which are particularly well preserved in partially-cemented silty beds. At depth, the calcareous shells are usually preserved, although many are broken, only the thicker-shelled bivalves and gastropods being preserved intact. The amount of shell material in the Crag, some 10 km east of Colchester, was commented on by Hollyer (1974, p.7) who noted that it is extremely variable and can reach 45 per cent by weight of the sediments. In contrast, the proportion of gravel in the Crag is usually less that 5 per cent by weight, and the clasts are generally less than 10 cm in diameter. They consist of well-rounded black flints with rounded quartz, vein-quartz and some subangular flints. Rounded black phosphatic nodules less than 2 cm in diameter are characteristic of the basal Crag. None was noted at outcrop but several of the nodules were recorded from boreholes in the north-east of the district.

Details

Crag 4.7 m thick was proved in a borehole near Farthing Hall [946 351]. In the outlier north of Bures, 4.5 m were proved whilst up to 3.6 m were penetrated in boreholes near Alphamstone, and 0.3 and 1.8 m near Wethersfield. Crag was proved in a BGS trial pit [9097 2521] near Aldham where less than 0.6 m overlies London Clay.

The only study of the palaeontology of the 'Red Crag' of the Braintree district was carried out on samples from between 21.1 and 22.9 m in a borehole put down near Shalford [7051 2974] (Clarke and Ambrose, 1975). The calcareous foraminifera obtained from the shelly fauna were: *Ammonia batavus*, *Bulmina elongata*, *Cassidulina laevigata*, *Cibicides lobatulus*, *Elphidiella hannai*, *Elphidium asklundi*, *Elphidium bartletti*, *Elphidium clavatum*, *Elphidium crispum*, *Globigerina bulloides*, *Oolina melo*, *Protelphidium* cf. *obiculare*, *Rosalina viladebeana*.

Grading curves have been obtained from samples of the 'Red Crag' from boreholes in this district for sand and gravel resource assessment surveys. They show a particle size distribution which is generally readily distinguishable from that of most drift sands and gravels (see Figures 20 and 21). Using these results, Hopson (1981), Marks and Merritt (1981) and Marks and Murray (1981) have discovered that non-shelly 'Red Crag' sands are perhaps more extensive in the northern part of this district than was recognised during the geological survey; and are probably represented locally in the lowest parts of the overlying sediments that have been ascribed to the Kesgrave Sands and Gravels. The possibility of recycling of the Crag sediments should, however, not be overlooked.   RAE

CHAPTER 5

Quaternary Drift: pre-Hoxnian

A division of the Quaternary into stages reflecting climatic oscillations and based largely on biostratigraphical studies in East Anglia was established by West in 1963. This subdivision has formed the framework for all subsequent stratigraphical studies and a later revision is tabulated below.

Correlation of the established British glacial and interglacial stages with those in continental Europe has been attempted on palaeontological and biostratigraphical grounds by many authors. The comments of Mitchell and others (1973) illustrate the problems involved in such correlation and make it clear that at present only the following tentative correlation can be made:

Britain	Europe
Flandrian	Holocene
Devensian	Weichselian
Ipswichian	Eemian
Wolstonian	Saale
Hoxnian	Holstein
Anglian	Saale/Elster

An alternative correlation put forward by Bristow and Cox (1973) and also mentioned in Mitchell and others (1973), suggested that the Wolstonian stage represents only a short-lived climatic deterioration within a major interglacial embracing both the Hoxnian and Ipswichian stages, which they suggested equate with the Eemian. Woodland (1970) took a more extreme view and suggested that the entire suite of glacial and post-glacial deposits of East Anglia (i.e. Anglian and post-Anglian) might be equivalent to the Weichselian of continental Europe.

A note of caution regarding attempted correlations of glacial and interglacial stages was sounded by Kukla (1977), who points out that the study of the oxygen isotope ratios and their variation with time will eventually establish a world-wide pattern of climatic change throughout the Quaternary. A consideration of the oxygen isotope data already available suggests that the simplistic tradition of the small number of alternating warm and cold episodes that has been erected from onshore stratigraphic studies do not necessarily tie in with the pattern of world-wide climatic fluctuations. In this context, the onshore sediments may have been deposited in only a fraction of the time represented by the offshore succession from which the oxygen isotope curves are obtained.

Although climatic fluctuations are the overriding consideration in the definition of the Quaternary stages, their number and relative superposition are, nevertheless, governed by stratigraphic principles applied as a result of examination of boreholes and sections over wide areas and can be verified in some cases only by regional geological mapping.

Taken together the post-Crag and pre-Hoxnian deposits form the greater proportion of the superficial sediments of the Braintree district. They are present over about 90 per cent of the district and attain a maximum thickness of 40 m. Three main 'units' are represented: Kesgrave Sands and Gravels, Glacial Sand and Gravel, and Boulder Clay (Figure 7). Their general stratigraphic relationship across the district is illustrated in Figure 8. The deposits have been deposited on a generally planar surface but dissecting this are several channels broadly coinciding with present-day river valleys. In the channels there is usually a complex inter-bedded sequence of Boulder Clay, Glacial Sand and Gravel, and Glacial Silt. The location of these channels with contours on bedrock beneath glacial deposits is shown in Figure 9.

The two sand and gravel formations and the Crag together constitute a single widespread sand and gravel body. In places all three units are present and separable in borehole logs, though only with difficulty on field evidence alone.

When the field survey commenced in 1969, only one sand and gravel unit was recognised. During 1974–75, however, excellent exposures in working gravel pits near Colchester and data from assessment boreholes enabled the recognition of two lithologically distinct sand and gravel units which could be individually mapped and which are here described as Kesgrave Sands and Gravels (lower unit) and Glacial Sand and Gravel (upper unit). At the same time, outliers of shelly Crag were recognised in the northern part of the

WOOD, 1867	PRESTWICH, 1890 SOLOMON, 1935	WHITAKER, 1878	CLAYTON, 1957	BRISTOW AND COX, 1973	MITCHELL, 1973	ROSE AND ALLEN, 1976, 1977	THIS MEMOIR
UPPER GLACIAL	CHALKY BOULDER CLAY	BOULDER CLAY	SPRINGFIELD TILL	SPRINGFIELD TILL (CHALKY BOULDER CLAY)	SPRINGFIELD TILL	LOWESTOFT TILL	BOULDER CLAY
MIDDLE GLACIAL	GLACIAL SAND AND GRAVEL	GLACIAL SAND AND GRAVEL	CHELMSFORD GRAVELS	CHELMSFORD GRAVELS	CHELMSFORD GRAVELS	BARHAM SANDS AND GRAVELS	GLACIAL SAND AND GRAVEL
	WESTLETON BEDS			(GLACIAL SAND AND GRAVEL)	WHITE BALLAST	KESGRAVE SANDS AND GRAVELS	KESGRAVE SANDS AND GRAVELS

Figure 7 Development of the terminology of the 'glacial' deposits

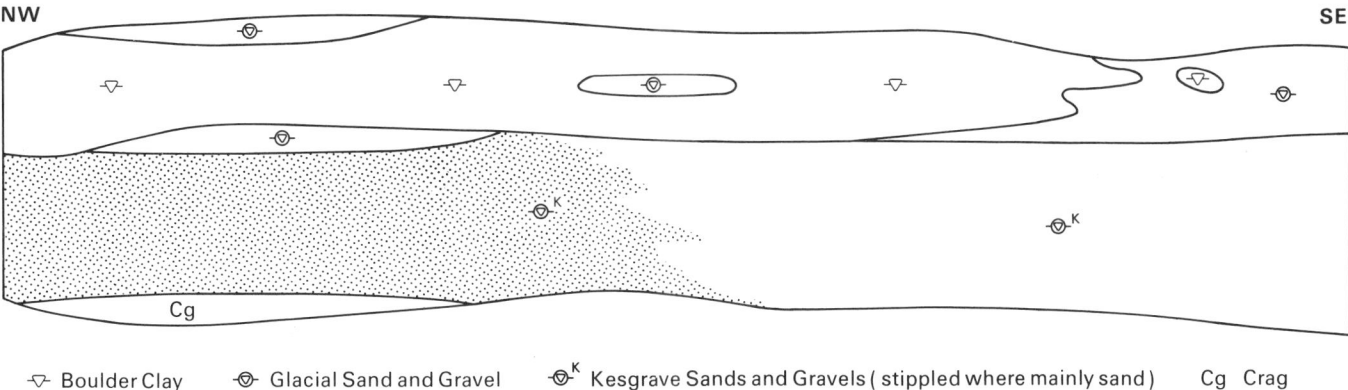

NW   SE

▽ Boulder Clay   ⊙ Glacial Sand and Gravel   ⊙ᴷ Kesgrave Sands and Gravels (stippled where mainly sand)   Cg Crag

Figure 8   Diagrammatic representation of the pre-Hoxnian Quaternary
deposits across the district

district, and subdrift occurrences of shelly Crag were
delineated on borehole evidence.

After completion of the field survey, further boreholes
were drilled in the northern part of the district, and the three
formations were widely identified in borehole samples by
pebble counts and grain-size distribution, although it had
not proved possible to separate them at outcrop. Their
characteristic grain size distribution curves are illustrated in
Figure 21. The Kesgrave Sands and Gravels shown on the
map may, therefore, include Glacial Sand and Gravel in
their upper part and Red Crag in their lower part (Hopson,
1981; Marks and Merritt, 1981; Marks and Murray, 1981).

The broad stratigraphical relationships of the glacial
deposits in the Braintree district have been known for over a
century. Wood (1867) recognised his Upper and Middle
Glacial deposits in this district (Figure 7); the Upper Glacial
was subsequently mapped by the early Geological Survey
officers as Boulder Clay and the Middle Glacial as Sand
and Gravel.

The Geological Survey classified all the sand and gravel
deposits in this district as 'Glacial'. Prestwich (1890), from
a study of quarry sections, cliff exposures and boreholes,
traced beds of 'pre-glacial' quartz-rich gravel from Norfolk
through Suffolk and into the Braintree district, where he
recorded them in cuttings exposed during construction of the
Sudbury to Marks Tey and Braintree to Witham railways.
He named them the Westleton Beds and regarded them as
marine in origin. He considered that they had not previously
been separated from Glacial Sand and Gravel mainly
because they were generally obscured by overlying glacial
deposits. Solomon (1935) agreed with Prestwich and clarified
the diagnostic characteristics of the Westleton Beds by study-
ing their heavy mineral assemblage. He noted that they contained Tertiary flint, quartz and some subangular flints and
Lower Greensand chert. In contrast the Glacial Sand and
Gravel was commonly 'dirty', and contained large boulders
of flint, quartzite, quartz and felsite, but no Jurassic
material. From a study of the deposits exposed in the Braintree district he concluded that most of the Glacial Sand and
Gravel as mapped there by the Geological Survey fell within
the Westleton Beds (see Figure 7).

No further original work on these sands and gravels was
undertaken in the Braintree district or adjacent areas until

Clayton (1957) described Glacial Sand and Gravel from the
Chelmsford and Harlow areas. Then from 1966 to 1970 the
primary six-inch geological survey of the Chelmsford (241)
district was carried out. During that survey the term Glacial
Sand and Gravel was retained, for no subdivision of the formation seemed possible. When the work was extended into
the Braintree district it became increasingly obvious that two
distinct suites of sand and gravel were present and, in the
eastern part of the district, it proved possible to map them
separately.

The distinction was aided by the work of Hey (*in* Rose and
Turner, 1973), who had revived the theories of Prestwich
and Solomon. In Alphamstone quarry [871 357] in this
district, he noted that the sand and gravel contained a high
proportion of quartz, quartzite and rounded quartz pebbles
consistent with Prestwich's Westleton Beds. Hey and Turner
(*in* Mitchell and others, 1973) both proposed that the deposit
was pre-Glacial (i.e. pre-Anglian) in age, and Rose, Allen
and Hey (1976) documented similar deposits, terming them
Kesgrave Sands and Gravels, with a type area in the Gipping
valley. They identified the formation in quarries within an
area of over 2000 km² from Ipswich to the Ongars, including
five localities in the Braintree district and others in the
Chelmsford district (Rose, Sturdy, Allen, and Whiteman,
1978): this term has been adopted in the present account.
The term Glacial Sand and Gravel is consequently used in a
more restricted sense than on previous Geological Survey
maps of East Anglia.

These two sand and gravel formations are described
separately below although in some areas it proved impossible
to distinguish the formations individually, even where
boreholes and quarry exposures have shown the two deposits
to be present. In such areas the two lithological types have
been grouped together for reasons of practicality, as
Kesgrave Sands and Gravels.

The boulder clay of this district has long been known to be
part of an extensive sheet extending across East Anglia
(Bristow and Cox, 1973; Perrin, Rose and Davies, 1979).
Harmer (1904) referred to the deposit as the product of the
Great Eastern Glacier, and qualified his remarks by noting
that the changing erratic content reflected the bedrock
lithologies over which the ice-sheet had passed. In East
Anglia a two-fold subdivision of the boulder clay was pro-

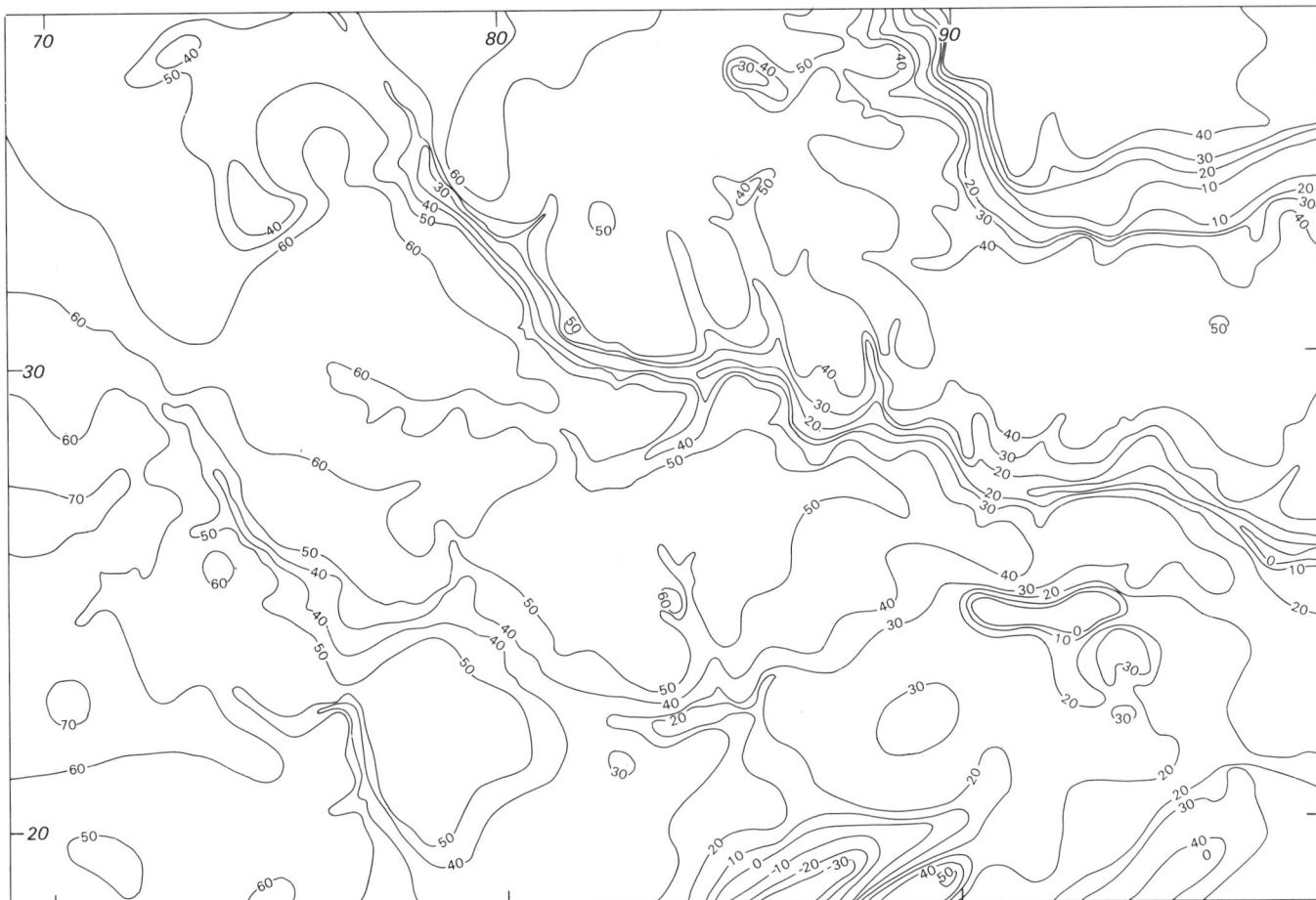

Figure 9 Rockhead contours at 10 m intervals

posed by Baden-Powell (1948), who noted that earlier authors had suggested that two chalky boulder clays might be present. He thought that the boulder clay with much Jurassic material, usually the lowest beds, should be separated from those with much Chalk. The former type he named the Lowestoft Boulder Clay, which he considered to have been deposited from ice moving in a south-eastward direction across Lincolnshire and into East Anglia; the latter type he named the Gipping Boulder Clay which, he argued, because it contained large amounts of Chalk, must have been derived from ice travelling along the strike of the Chalk outcrop. In 1956, West and Donner studied the orientation of the long axes of pebbles in the boulder clay in an attempt to confirm the direction of ice movement. They concluded that in the Braintree district and adjacent areas the Lowestoft till was deposited from an ice-sheet moving from the north-west, whilst the Gipping till advance was from a more northerly direction. Since then, however, it has been generally agreed by most workers (Bristow and Cox, 1973; Perrin, Davies and Fysh, 1973; and Perrin, Rose and Davies, 1979) that only one extensive chalky boulder clay is present in southern East Anglia, and that this was laid down during a single glacial episode. This conclusion is supported by the results of this survey of the Braintree district.

It is now universally accepted that a considerable time interval separated the deposition of the Crag and the succeeding Kesgrave Sands and Gravels. Differing views of the depositional history of the latter have been presented by various authors (Baker and Jones, 1980; Rose, Allen and Hey, 1976; Hey, 1980). It is now generally accepted that they are 'proto-Thames' river deposits laid down by an eastward-flowing braided river. Their precise age in terms of the Quaternary stages remains disputed but, in the absence of faunal evidence, is inferred to be either Anglian (Baker and Jones, 1980) or Beestonian (Rose, Allen and Hey, 1976). These interpretations of the age depend heavily on the dating of the proto-Thames river diversions to the west of this district (see Millward and others, *in press*, and Baker and Jones, 1980). The Kesgrave Sands and Gravels in this district vary in lithology from cross-bedded sands with large scale cross-sets in the north, to planar-bedded more gravelly sediments with relatively small cross-sets in the south and east. The areas with these contrasting lithologies are separated by a tract, running across the central part of the district (see Figure 17), where gravels are thin or absent. The lithological variation reflects a change in energy conditions and sediment supply to the river system, which was gradually migrating southward.

Following the deposition of the Kesgrave Sands and Gravels, relatively stable conditions allowed the formation of a widespread soil. The 'B' horizon of the resultant fossilised soil profile, where preserved, is reddened, a feature attributed to a temperate climate (Rose, Allen and Hey, 1976). Subsequently the profile was contorted by ground ice during the onset of a cold period.

During this cold period (Anglian stage) a major ice-sheet advanced across most of northern and central Britain. In the Braintree district this ice laid down an outwash of Glacial Sand and Gravel and deposited a lodgement till, the chalky Boulder Clay, over this outwash. The mechanism of the ice advance is little understood, but it is apparent that during the advance, subglacial streams under considerable hydrostatic pressure excavated deep, steep-sided escape routes and tunnel valleys (Woodland, 1970). Most of these valleys were cut into the Chalk bedrock parallel to the direction of ice movement, and in many instances were incised to depths greater than 50 m below the regional bedrock surface. As these glacial valleys cross the Chalk subcrop on to Tertiary strata they become considerably reduced in magnitude, suggesting that it was the hydrostatic head in the phreatic Chalk water which accounted for a large proportion of the erosive power that created the tunnel valleys. Outside these 'active zones' the sole of the ice-sheet appears to have been well lubricated and there is little evidence of erosion, a fact confirmed by the widespread preservation of the pre-glacial soil. At its maximum limit, the ice-sheet reached the southern part of the Braintree district where it was impeded by a London Clay ridge stretching from Birch to Tiptree and beyond. The present day drainage pattern was probably initiated during this glacial period, the present major valleys coinciding with the main glacial outwash routes. During the waning glaciation these active meltwater channels became choked with outwash deposits consisting of Glacial Sand and Gravel or Glacial Silt (in temporary ice-dammed lakes) together with some Boulder Clay. On the interfluves, the ice-sheet melted to leave a till, underlain in part by thin Glacial Sand and Gravel.

KESGRAVE SANDS AND GRAVELS

The formation is widely spread in the district. It has a sheet-like form and invariably lies beneath the Boulder Clay. It crops out on the valley slopes of the major rivers and many of their tributaries, and locally has a distinctive topographical expression. Where more that 10 m of Kesgrave Sands and Gravels are present (that is, in general terms, the River Stour valley, the River Colne valley north of Halstead and in the Roman River valley) they crop out on steep convex spurs with a pronounced break of slope at their base, associated with vigorous springs emanating on the underlying London Clay. As the result of extensive field draining schemes, the springs have now been intercepted and most are confined to the heads of the tributary valleys.

The areas where Kesgrave Sands and Gravels are probably absent have been assessed from borehole evidence (Figure 10). Broadly, these areas (Clarke and Ambrose, 1975; Booth and Merritt, 1982) are: between Wethersfield [71 31] and Toppesfield [74 37]; from Bannister Green [70 20] through Rayne [73 26] to Panfield [74 25] and High Garrett [78 27]; around Cressing [79 21]; in the Janke's Green area [905 300]; and between Coggeshall [86 23], Pott's Green [91 23] and Old Will's Farm, Kelvedon [88 21].

Field mapping has permitted a general appreciation of the lithological variation of the unit, particularly where it is thick and where augering has been augmented by small exposures, disused sand and gravel pits and more extensive active workings. This information has been added to by a systematic programme of borehole drilling undertaken as part of a regional sand and gravel assessment programme funded by the Department of the Environment: the most recent boreholes in the northern part of the district (Hopson, 1981; Marks and Murray 1981) have proved to be particularly useful because they also include pebble counts.

The base of the Kesgrave Sands and Gravels over most of this district is sharply unconformable on Tertiary strata or Chalk. Locally, where Crag intervenes, the base is less easily defined, both in boreholes and at outcrop. In such places the base has been drawn immediately above shelly Crag. A further description of the distinction between Crag and Kesgrave Sands and Gravels appears later in this account (p. 56).

Although the deposit is variable, two main interdigitating lithologies of sands and gravelly sands have been recognised. In an area bounded to the south approximately by Shalford [720 290], Gosfield [784 295], Sible Hedingham [782 340] and Wickham St Paul [831 364], the Kesgrave Sands and Gravels consist almost exclusively of well sorted medium-grained quartz sand. The northern limit of these sands probably lies beneath a thick Boulder Clay cover in the north-west of the district although borehole information in that region is sparse and no firm evidence is available. Sand beds of similar lithology also occur throughout the Kesgrave Sands and Gravels in the remainder of the district; they are thickest in the basal part in the north, and interbedded with gravel in the south. During the present survey, sands were well exposed in a working [722 287] at Shalford, in another near Beazley End [737 287], at Brook Street Farm, Sible Hedingham [798 319] and Purlshill [793 339].

The sands attain a maximum thickness of over 20 m in the area of Blackmore End [739 309]; they thin southward and eastwards where there are increasing numbers of gravelly interbeds in the upper part of the Kesgrave Sands and Gravels. The sand beds have been recorded in the vicinity of Braintree by French (1891) as ferruginous or 'rich yellow coloured' micaceous and 'white.... silver sand', and by Solomon (1935) around Colne Engaine [851 305].

In more detail, the sand grains are subrounded quartz clasts of medium grade (0.25 to 0.5 mm diameter) (see Figure 21), and are transparent or have a thin iron oxide coating which imparts an overall rusty brown or orange colour to the sediments (10R to 5YR hues); where the concentration of oxides is low, the deposit is white or pale yellow (10YR hues). Routine grain size analyses of samples from the lower part of the sands has shown them to possess grading characteristics similar to those of the Crag (see Figure 21; Hopson, (1981) and Marks and Murray, (1981)), and these sands have been separated from the Kesgrave Sands and Gravels and considered to be Crag. Muscovite flakes up to 2 mm in diameter are common on bedding surfaces and black disseminated iron oxide occurs as small powdery grains which might be mistaken for carbonaceous material. Beds and lenses, generally less than 20 cm thick, of pale greenish grey micaceous laminated clayey silt and clayey fine sand occur within the sands. These argillaceous horizons and the amounts of mica decrease south and

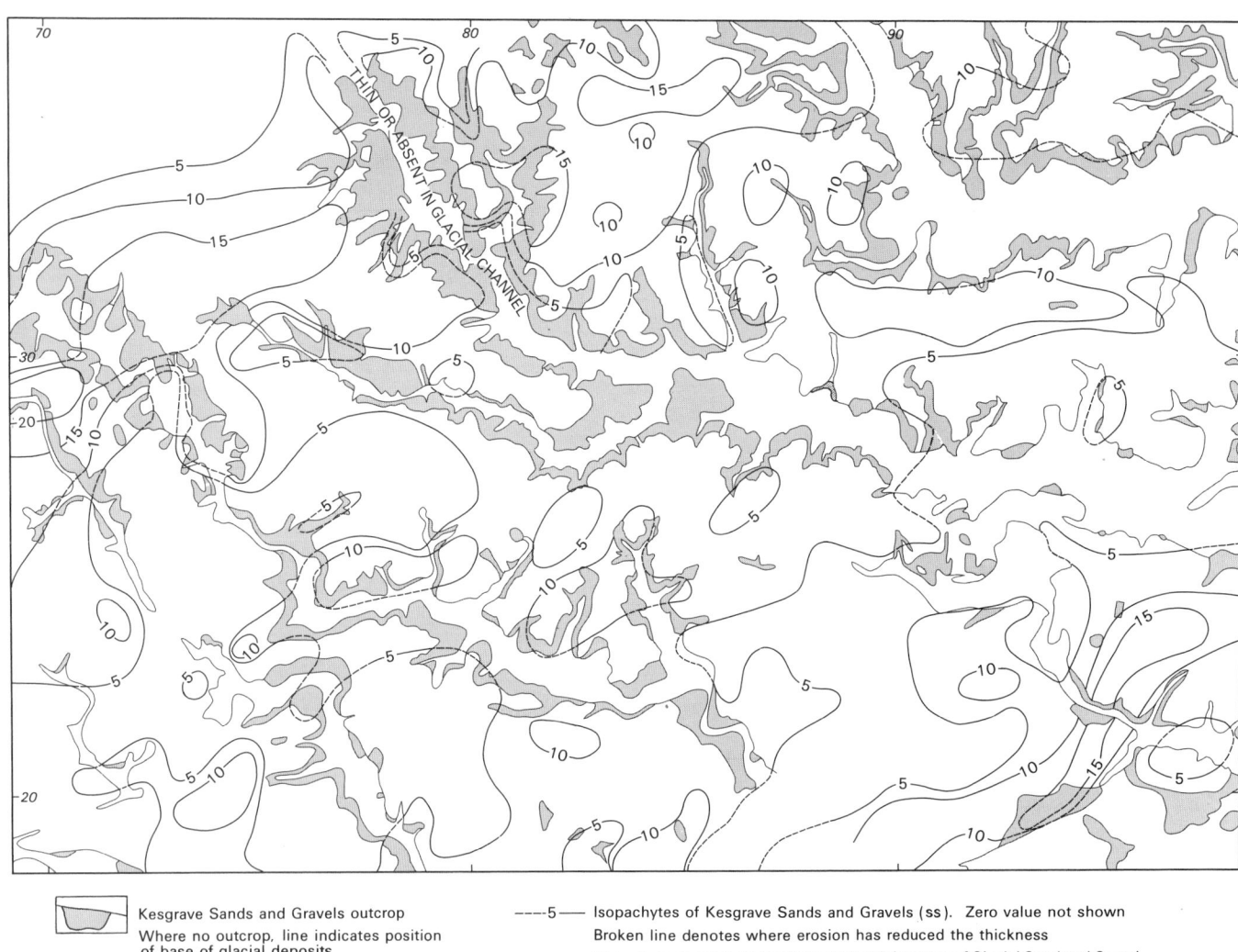

Kesgrave Sands and Gravels outcrop
Where no outcrop, line indicates position
of base of glacial deposits

----5---- Isopachytes of Kesgrave Sands and Gravels (ss). Zero value not shown
Broken line denotes where erosion has reduced the thickness
The values shown here locally include thicknesses of Glacial Sand and Gravel
(the upper clayey beds) and of non-shelly Crag

Figure 10   Isopachytes of the Kesgrave Sands and Gravels

eastwards across the district. Similarly well sorted sands with green clay beds which occur north-east of this district (Allender and Hollyer, 1972) have been equated with the Chillesford Beds of north-east Suffolk. In the Braintree district, Clarke and Ambrose (1975) have tentatively suggested the same correlation although, in the absence of firm palaeontological or mineralogical control, this suggestion has not been confirmed.

Gravelly beds within the Kesgrave Sands and Gravels predominate in the south and east of the district. They occur in five main areas (see also Figures 11 and 17):

1   In the vicinity of Youngs End [740 195], near Black Notley
2   Between Silver End [810 197] and Bradwell [818 221]
3   Between Pattiswick Green [820 250] and Markshall [841 255], north-west of Coggeshall
4   From Pebmarsh [854 335] to Wormingford [935 315], and north of the River Stour
5   A broad tract between Messing [897 189] and Stanway Green [960 235]

They are thickest in the last area, where over 18 m are present locally. Their colour varies depending upon the iron ox-

ide content, so that they vary from almost white (10YR hues) to orange brown (5YR) and deep rusty brown (10R hues). Generally the relatively iron-free layers occupy the middle part of the sequence which is known locally as 'Essex White Ballast'. During the present survey these gravelly beds were best exposed in quarries at Alphamstone [849 354] (closed in 1976), Ferriers [896 342], Beazley End [737 287], Bradwell [820 217], Stanway [949 236] and Birch [926 200].

The gravel clasts are mostly rounded, and the deposit consists largely of rounded, subrounded, and subangular flint pebbles with about 30 per cent rounded white quartz and purple, red or bleached white quartzite; other erratics form about 1 to 4 per cent of the gravel and these include durable sedimentary rocks such as chert, and igneous and metamorphic rocks. The commonest and most conspicuous igneous rock is a greenish coloured fine-grained acidic or felsitic rock whose pebbles range from 50 cm to less than 1 cm in diameter, although pebbles less than 5 cm are uncommon.

Several sections in railway cuttings between Braintree and Witham were examined by Prestwich (1890) and French (1891) who noted the distinctive pebble suite of the exposed sand and gravel beds. Salter (1905) examined the exotic

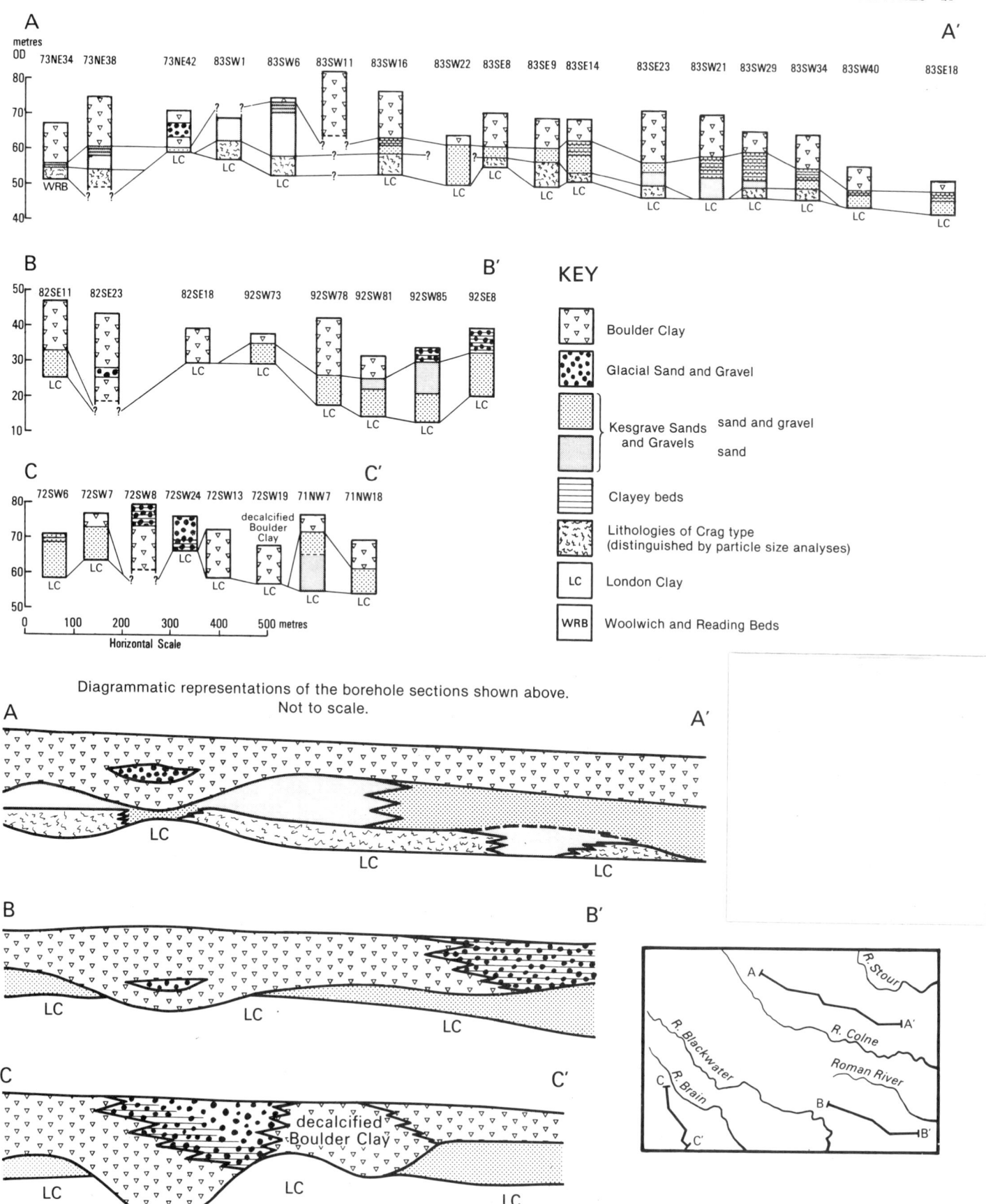

Figure 11 Correlation of drift deposits

pebbles in more detail and listed 'Lower Greensand chert, radiolarian chert, jasper, Carboniferous chert, weathered igneous rocks like rhyolite, and greenish pebbles containing tourmaline'. The relatively high proportion of 'Bunter' and quartz pebbles was stressed by Solomon (1935). More recently, Hey (in Rose and Turner, 1973) has published a representative pebble count of the 16 to 32 mm gravel fraction of the Kesgrave Sands and Gravels in Alphamstone quarry (the section was obscured in 1979). He recorded 29 per cent rounded flint, 38 per cent angular and subangular flint, 16.5 per cent quartz, 13 per cent quartzite, 3 per cent chert and 0.5 per cent volcanic rocks. The chert fragments are grey and spicular and are thought by Hey to be derived from the Carboniferous. Pebble counts undertaken during the present survey on the 16 to 64 m clasts indicate an average of 10 per cent quartz, 13 per cent quartzite, 3 per cent sandstone, 45 per cent angular flint, 27 per cent rounded flint and 1 per cent igneous rocks; in the 4 to 16 mm fraction quartz makes up 26 per cent of the clasts and the proportion of flint is smaller.

The heavy mineral content of the Kesgrave Sands and Gravels was examined by Solomon (1935) who recognised two assemblages: the basal part characterised by tourmaline with subsidiary amounts of staurolite, zircon and rutile; the upper part by amphiboles, along with garnet, epidote, apatite and sphene.

The formation exhibits a wide variety of sedimentary structures. In the northern part of the district, large-scale cross-stratification occurs with cross-sets up to 3 m amplitude in the sand beds. The beds are finely laminated and small-scale ripples are also present in the top part of the cross-sets. The foresets dip between north-east and south-east. In more southern parts, in particular in quarries at Stanway, Bradwell and Silver End, the large-scale cross-stratification is not so well displayed but is replaced by planar bedding within which are low amplitude (less than 1 m) cross-sets commonly with ripples and ripple drift lamination in the finer grained sandy beds. There are, however, larger cross-sets in the bottom part of the sequence at Stanway and Birch.

Deformation structures are common in the Kesgrave Sands and Gravels, as a result of both contemporaneous and post-depositional periglacial activity. In the quarries at Warren Lane, Stanway [950 225] and at Silver End [816 191], there are small-scale London Clay linear diapiric structures penetrating 1 m or more into the overlying sand and gravel. Although no clear sections were available during the present survey, the width of these superficial injections of London Clay is probably less than 2 m at their roots but they are at least 10 m long and may reach 30 m. Similar, but better exposed structures, have recently been described from the adjacent Epping (240) district (Millward and others, in press) where they form polygonal patterns. In the Stanway White House quarry [949 237], a London Clay diapir or possibly a pinnacle, extends vertically through at least 10 m of sand and gravel and is injected along a bedding plane in the sand and gravel for a distance of 150 m. London Clay clasts were noted in the sand and gravel above the clay injection proving this to have been contemporaneous with the deposition of the sand and gravels. The mechanism of such failure and flowage of London Clay is little understood. The initiation of the injections was probably induced by excess pore pressure

in the London Clay which came about on melting of ground ice. With rapid deposition of the Kesgrave Sands and Gravels, the high pore-pressures were presumably maintained, and mobilisation and injection of the London Clay followed.

Conical-shaped disturbances in the Kesgrave Sands and Gravels were noted in the Bradwell and White House quarries at Stanway. These are ice-wedge casts and they vary from less than 1 m to over 3 m in diameter at their upper limit and extend through up to 5 m of sediments. They have been generally filled with silts and poorly sorted sand and gravel, whose bedding has been downturned into the wedges (see Plate 2); sympathetic normal step faulting with throws of up to 60 cm are common in the sands and gravels adjacent to the wedge. Similar sub-vertical normal faults with throws not exceeding 30 cm occur in the Kesgrave Sands and Gravels even where ice-wedge casts are not evident.

Further bedding interruptions in the form of lobate involutions occur in the upper part of the formation. They have been observed at Beazley End, Ferriers, and Silver End quarries, where they are associated with lithologies considerably more clayey and silty than those of the underlying beds, and with the development of orange brown (5YR hues) and red (5R hues) mottling. The involutions are thought to be the result of periglacial freeeze-thaw activity associated with ground ice. Rose, Allen and Hey (1976) and Rose and Allen (1977) have interpreted such features as an arctic structure soil superposed on the red illuvial horizon of a temperate soil which they have recognised widely at the top of the Kesgrave Sands and Gravels in southern East Anglia. The discovery of such a soil horizon in a drift sequence has important stratigraphical consequences in that it establishes a chronostratigraphical datum. It was not possible to determine the extent of the involuted horizon during the present survey, but many boreholes penetrated the typical red brown clayey and silty sands and gravels which are normally associated with it. These clayey beds are generally 1 to 2 m thick but in places 5 m are recorded.   RAE

Details

Western sheet boundary to Pods Brook and the River Brain

Within this area the Kesgrave Sands and Gravels have an irregular distribution (see Figure 10). In several places there is no sand and gravel either at outcrop or at depth, and the absence coincides in general with a marked increase in thickness in the overlying Boulder Clay (see Figure 13).

North of Great Saling area [c. 705 265] it is possible to divide the Kesgrave Sands and Gravels into a lower, almost entirely sandy part, and an upper gravelly part. This two-fold division has been recognised in several boreholes farther north and for some distance to the north-east (see p. 23).

Around Rumley Wood [710 238], sand and gravel, varying from 5.5 to 10.1 m in thickness, is present at depth, and as a small outcrop in the vicinity of Blake End [702 228]. The presence of significant amounts of gravel throughout the deposit suggests that it may correspond only to the upper part of the Kesgrave Sands and Gravels recognised farther north.

South-east of a line from Thistley Green [706 192] to Stanford Farm [734 220], the sand and gravel is present at depth; up to 21.6 m has been proved, although in general the thickness averages about 7 m.

Plate 2   Ice wedge cast in Kesgrave Sands and Gravels in the Warren Lane pit, Stanway

Kesgrave Sands and Gravels crop out continuously along the western side of the Brain valley from Braintree to the margin of the district. The deposits on the eastern side are not so continuous, and in places Boulder Clay rests directly on the London Clay. Springs issuing from the base of the sand and gravel are common.   CRB

Pods Brook to the River Pant

Thick deposits of Kesgrave Sands and Gravels (up to 23 m) are present in the Shalford area (see Figure 10) whereas the deposit is thin or absent towards Panfield and Braintree. This thinning apparently takes place gradually although there are local anomalies, as at Iron Bridge Farm [728 282], which are probably due to channelling at the base of the Boulder Clay. The sediments dominantly consist of medium- to fine-grained micaceous sands; sandy gravels overlie these beds locally as, for example, in a borehole [7031 2871] near Park Hall (Clarke and Ambrose, 1975).

North-west of Shalford Pond Green [711 272], Whitaker and others (1878, p.44) recorded 3.7 m of 'yellow sand with thin bands of brown hard sand'. He also noted two minor exposures at Jasper's Green [720 267] in sand and gravel, presumably lying beneath thin Boulder Clay.

At the disused Shalford pit [722 285], up to 2 m of crudely bedded, ferruginous sandy gravel with a dominantly clayey sand matrix rests with a sharp channelled base on orange-yellow, cross-bedded, fine to medium-grained micaceous sands. The lower sands

showed well-defined trough-bedding (with troughs up to 0.5 m deep) and microfaults, and were exposed for up to 5 m. Excavations in sands were noted by Whitaker and others (1878, p.45) at this locality and near Abbot's Hall [7335 2780].

The old gravel workings at Bradwell [818 217] showed:

	Thickness m
Chalky boulder clay with sandy horizons, grey but decalcified; gravelly and brown in the basal 0.5 m; calcrete 0.07 m thick at base	3.5 to 4.5
Sands and gravels, ochreous with clayey, silty and fine-grained sandy horizons; cross-bedded; ripple structures	1.8 to 2.5
Sand and gravel, well-bedded, white, coarser than above	2.7 to 3.5

At the north end of the pit, where the Boulder Clay is absent, there is an ice-wedge structure in the gravels beneath Head.   RDL, RAE

River Pant to the River Colne

In this area boreholes show that the deposits are sheet-like in form, and generally between 5 and 15 m thick. They are locally absent near Coggeshall, to the east of Gosfield, and around High Garrett (see Figure 10). There is no information in the Toppesfield area.

Between Shalford and Sible Hedingham, where the Kesgrave

Sands and Gravels are thickest, the following sequence is typically present beneath the Boulder Clay:

	Thickness m
Sandy gravels, with flints and quartzite pebbles; clayey at the top; exceptionally 6 m thick, locally absent, but more usually	2.5 to 4
Sands, fine- to medium-grained, micaceous, with clay partings; local pebbly partings; generally makes up the bulk of the sequence	up to 20
Sands, as above, tending to coarsen downwards, with pebbly partings; locally there is a separate gravel unit with quartz and flint pebbles; exceptionally up to 9 m thick	0 to 3.5

The lower sands show a coarsening of grain-size downwards possibly due to incorporation of locally-derived Crag. Whitaker and others (1878, p. 47) noted that the fullest sequence of beds beneath the main Boulder Clay sheet in the Sible Hedingham area is: fine-grained sand – lenticular boulder clay – quartzose pebbly sand in ascending order. Comparison with the above sequence suggests that the 'lenticular boulder clay', which was not located in the recent survey, may be a local flow-till beneath the topmost sandy gravels.

The Beazley End [737 293] gravel-pit exposed the following sequence in 1977:

	Thickness m
Chalky Boulder Clay	5.0
Gravelly sandy clay, distinctly red and orange-brown mottled, involuted; involutions incorporate red and grey colour-laminated beds with dominantly rounded pebbles about 2 cm diameter; indistinct base	1.5
Sands, fine-grained, orange-brown, patchily mottled pale grey, locally cross-bedded; passing laterally into silty sands with pebbly partings	up to 2.0
Sands, pale grey, cross-bedded	up to 10.0

The workings [736 291], now backfilled, were said to have exposed, below the Boulder Clay, poorly-bedded ferruginous sandy gravels which rested with a channelled base on the cross-bedded sands.

To the south of Codham Mill [7355 2817], the disposition of the drift outcrops indicates that the Boulder Clay has a channelled base along the valley-axis, and the Kesgrave Sands and Gravels are completely eroded away in places. A borehole [7627 2535], sited near the axis of the channel at Doreward's Hall, Bocking, proved:

	Thickness m
Chalky Boulder Clay	8.8
Soft silty clay (?lake deposits)	2.4
Gravel and silty clay	3.4
London Clay	0.9

The gravel pit at Strait's Mill, Bocking showed the following succession [7787 2460] in 1977:

	Thickness m
Sands, medium-grained, ochreous with silt partings and subordinate pebbly partings; cross-bedded	3.0
Sandy gravels, clayey, crudely bedded, variably sorted; angular and rounded pebbles; large irregular flint cobbles at base	1.5
Chalky Boulder Clay, greyish brown, weakly laminated; pebbles are angular and rounded and lie with a preferred horizontal orientation (flow-till); slightly decalcified; sharp transgressive base	0 to 1.5

	Thickness m
Sands and gravels, bedded, white to pale grey; dominantly rounded pebbles	seen to 3.0

This section is notable for showing the presence of a possible flow-till within the sands and gravels. A disused face [7757 2461] farther to the west showed the beds which apparently lie below the chalky Boulder Clay:

	Thickness m
Clay, sandy, silty with scattered dominantly rounded pebbles, red-brown and grey mottled	1.5
Gravels, with a clayey coarse sand matrix, reddish brown, cryoturbated	up to 1.5
Gravelly sands, bedded	1.5

The Foxborough Hill gravel pit [795 324] showed the following sequence:

	Thickness m
Chalky Boulder Clay, grey; paler in colour at the base	2.5
Clayey sands and sandy clays, colour-laminated in red, grey and brown; pockets of rounded pebbles to 2 cm diameter scattered throughout	1.5
Involuted beds, lithology as above, with local relict colour-lamination; pebble content increases downwards	1.5
Sands, with subordinate gravel interbeds, dominantly white, cross-bedded; local clay partings	4.0
Sands, medium- to coarse-grained, cross-bedded; local clay partings	3.0
Sand, fine- to medium-grained, micaceous	seen to 1.0

A lens of pale grey silty clay, up to 0.3 m thick, possibly of lacustrine origin, was noted beneath the involuted beds which were present along the face for several hundred metres.

Boreholes to the north and east of Bradwell [8424 2273; 8348 2353; 8288 2531] proved laminated micaceous silts at the base of the sand and gravel.

In a pit near Hill Farm, 0.4 km north of Coggeshall Church, Whitaker and others (1878, p.45) noted sands with 'some layers of dark reddish brown clay like weathered London Clay'. This location cannot be identified but the deposits are probably solifluction deposits.

RDL

River Colne to the Roman River

Eastward from Chappel [890 278], sand and gravel crops out on the southern side of the River Colne valley as far as 250 m east of the railway [9010 2685]. Yellow coarse sand was augered [8897 2756], while coarse sand and gravel crops out in a road cutting [8925 2755]; a borehole close to the cutting commenced in Kesgrave Sands and Gravels and proved a further 5.6 m overlying London Clay at + 42.3 m OD. Sand and gravel also crops out in a railway cutting [898 266] beneath about 6 m of Boulder Clay; east of the railway the sand and gravel is attenuated and Boulder Clay rests directly on London Clay at around 46 m OD [9018 2674]. The Boulder Clay thins southwards across the plateau and has been removed by erosion in places exposing the Kesgrave Sands and Gravels beneath [9080 2580; 9110 2580]. Extensive tracts of sand and gravel have been mapped around Church House Farm [9072 2531] and smaller areas elsewhere [916 251; 915 248]. In these areas the soils are particularly gravelly, and medium-grained sand, which is clayey in places, was augered. A BGS trial pit [9150 2580], 250 m west of Aldham Church, exposed:

	Thickness m
Top soil	0.3
Head:	
Orange and grey mottled variably clayey medium to coarse sand with gravel; structureless; well defined cryoturbated base	0.7
?Head:	
Orange with grey sub-vertical veining, clayey medium to coarse sand; some subangular flints and rounded quartzite pebbles; structureless; sharp base	1.3
Kesgrave Sands and Gravels:	
Orange with grey mottling (as above), cross-bedded medium to coarse sand; slightly clayey in places; discontinuous lenses of subangular to subrounded flints and a few cobble-sized quartzites and sarsens	1.0
Interbedded yellow brown micaceous fine-grained sand and clayey coarse sand with fine to medium gravel	seen to 0.5

The structureless sand in the upper part of this pit may be Glacial Sand and Gravel which was not mapped in this vicinity. In this area the base of the Kesgrave Sands and Gravels appears irregular; it is at about 40 m OD in a ditch section [9060 2522], but drops steeply to 30.5 m OD in a borehole at Aldham Hall [9175 2487].

Shallow disused gravel pits lie 200 m south-west of Aldham Hall, and a BGS trial pit dug in the tract of gravel proved:

	Thickness m
Head	
Clayey sand and gravel	2.4
Kesgrave Sands and Gravels	
Greyish yellow fine to coarse sand and mostly rounded gravel from 2 to 3 cm diameter; laminated	seen to 1.1

It is likely that the workings extracted gravel from the Head deposits as well as from the Kesgrave Sands and Gravels.

Kesgrave Sands and Gravels occur only in irregular pockets in the area from Gallows Green [923 263] to Eight Ash Green [945 254]. Abundant angular to rounded gravel was noted on the spur [9364 2690] north of Porters Farms and old shallow gravel diggings were seen nearby [9345 2685]; the base of the Kesgrave Sands and Gravels here is at around 35 m OD. A borehole [9362 2675] sited on Glacial Sand and Gravel, 100 m upslope from the Kesgrave Sands and Gravels outcrop, proved 7.9 m of clay and clayey sand and gravel, resting on London Clay at 33.5 m OD. Other boreholes in this area proved 2 to 3 m of Kesgrave Sands and Gravels though one [9458 2537], near Bridge Farm, penetrated 4.6 m of reddish brown sandy gravel beneath Glacial Sand and Gravel.

A trench section at Newlands Farm near Eight Ash Green [947 259] showed 2 m of orange clayey sand passing into medium to coarse sand. The spur 300 m to the east has very gravelly soil which includes vein-quartz cobbles; a spring in the vicinity indicated that sand and gravel rests on London Clay at shallow depth beneath Head deposits [9485 2588].

From 2 to 7 m of sand and gravel occurs [938 248] beneath Glacial Sand and Gravel, 600 m east of Moat Farm.

During the present survey Kesgrave Sands and Gravels were well exposed in working quarries on both sides of Warren Lane, Stanway, where they reach a maximum proved thickness of 16.8 m and thin south-eastwards towards the Roman River valley. In general terms, 10 to 15 m are probably consistently present between Warren Lane and Colchester. The base of the Kesgrave Sands and Gravels falls from around 30 to 35 m OD between Eight Ash Green and Beacon End [954 248] to 13 m OD in a borehole [9468 2203] near the Roman River valley north of Heckfordbridge. This area, in which the Stanway quarries lie, coincides with Block F of

Ambrose (1974) in which the Kesgrave Sands and Gravels are appreciably finer grained than in other areas on Sheet TL 92; the highest percentage of gravel (i.e. particles greater than 4 mm diameter) in Block F is 34 per cent but the mean is only 23 per cent.

The ARC Stanway Bellhouse Farm pit [948 235] exposes up to 15 m of yellow and orange well-sorted fine to coarse sand and gravel, with some less well sorted rusty orange brown beds which have a silty matrix. The deposits are, in general, a sequence of planar bedded sand and gravel, but closer inspection reveals some cross stratification and trough-bedding, the latter incised up to 50 cm particularly in the lowest 10 m of the deposit. The foresets generally dip in a south-easterly direction. Some cross-bedded units, up to 30 cm thick, of gravel-free medium to fine sand occur; the sand mostly consists of subrounded opaque quartz grains. Pebble counts from the sorted gravel in the quarry stockpiles give a rough indication of the percentages of the various pebble types present in the Kesgrave Sands and Gravels: 70 to 80 per cent of the gravel is flint, of which 50 per cent is rounded to subrounded and 50 per cent subangular; 5 to 10 per cent is vein-quartz; 5 to 10 per cent is quartzite: vein-quartz constitutes a higher proportion (up to 20 per cent) of the pebbles less than 1 cm in diameter. Of the pebbles less than 5 cm diameter, up to 10 per cent are highly weathered and pitted sandstones of indeterminate provenance. Various other erratics including green igneous rock (tuff), weathered granite, dolerite, sandstone, Carboniferous limestone and Jurassic limestone, constitute about 1 to 2 per cent of the gravel. In the northern part of the quarry [9475 2386] a step-like feature in the London Clay surface was exposed beneath the sand and gravel; the bedrock surface rose northwards by about 5 m with a 20 to 30° slope striking approximately east-north-east. The thickness of the Kesgrave Sands and Gravels is accordingly reduced in the north face of the quarry, although no disturbance which could be attributable to the London Clay surface irregularity was observed in the sedimentary structures of the gravel. Nearby, however, in the west face, tongues and flames of a soft smooth brown silty clay about 4 m from the top of the Kesgrave Sands and Gravels are associated with minor thrusting which has resulted in distorted bedding.

In the upper part of the face [9510 2348] in the eastern portion of the quarry an excellent example of a frost-wedge was exposed, disturbing some 2 m thickness of strata. Small-scale normal compensation faulting was well displayed on its flanks (see Plate 2).

The Kesgrave Sands and Gravels in the Warren Lane Stanway quarry, owned by Francis Aggregates, showed essentially the same characteristics as in the ARC quarry. The thickness increases from south to north from 10 to 15 m, mirroring the thickness variation in the overlying Glacial Sands and Gravel. Ripple drift lamination within yellow-orange fine- to medium-grained sand beds was quite common in the middle to lower part of the western face [9470 2272]. In the southern end of the pit [9458 2251], where the Glacial Sand and Gravel is thickest (see p. 31), the Kesgrave Sands and Gravels are reduced to about 9 m in thickness.

The Warren Lane, Stanway (Hoveringham) sand and gravel quarry was poorly exposed during the survey, but excavations in 1976 [9513 2280] proved about 15 m of planar bedded sand and gravel, similar to the deposits in the pits described above, overlying London Clay.

A disused quarry at Shrub Lane [968 232] exposed about 10 m of Kesgrave Sands and Gravels whose base was not exposed. Evidence from a nearby water well indicates that a further 3 m of sand and gravel is probably present in this area overlying London Clay. In the pit, the Kesgrave Sands and Gravels are generally planar bedded; cross-stratification within the beds exhibits foresets dipping from the north-east. Individual beds are rarely greater than 1 m thick and comprise a sequence of orange medium to coarse sands with varying gravel content, and subordinate slightly iron-cemented medium to coarse sand which is particularly common in the upper part. Some shallow cross-bedded channel-fills up to 40 cm deep occur. The pebble content is similar to that in the quarries described above, the gravel consisting of about 20 per cent vein-quartz, 5 per cent quartzite, 40 to 50 per cent subangular flint, and 35 per cent rounded and subrounded flints (including well rounded black Tertiary-derived flints); secondary alteration of the flints varies considerably but generally the finer grade subangular flints (< 5 mm diameter) tend to be patinated.

In the Roman River valley, outcrops of Kesgrave Sands and Gravels occur on the steeper slopes. Springs mark the base of the deposit in the bottom of the river valley and along its tributaries.

Old sand diggings were noted in the bottom of an unnamed tributary valley running towards Shrub End [9594 2180].   RAE

Roman River to River Blackwater and southern district boundary

Kesgrave Sands and Gravels between 6 and 8 m in thickness (see Figure 10) lie beneath Boulder Clay between Coggeshall and Feering. The deposits crop out in the Blackwater valley, notably west of Feeringbury [8643 2155], and small disused gravel pits were noted nearby [8565 2167].

At Marks Tey, temporary excavations during construction of the A12 by-pass were visited in 1969 by Dr C. R. Bristow, who recorded the 'Essex White Ballast' type of Kesgrave Sands and Gravels in the floor of the cutting [9199 2402]. He also noted 2 m of gravel exposed beneath 4.5 m of chalky Boulder Clay near the A12 underpass [9179 2390]. Boreholes in this vicinity have proved a total thickness of Kesgrave Sands and Gravels of around 8 m, overlying London Clay. A borrowpit [924 240] near the Marks Tey Hotel was opened in 1969 and exposed about 6 m of white sand and gravel beneath Boulder Clay. A scapula fragment of *Bos*, an *Elephas antiquus* molar and a juvenile *Mammuthus primigenius* molar were recovered from this borrowpit and they are now registered in the Colchester Natural History Museum. By the time of the present survey the pit had been backfilled and no section was visible.

Extensive deposits of Kesgrave Sands and Gravels are present southwards from Whitehouse Farm [921 212] to the London Clay ridge running through Smyths Green [921 185] (see Figure 10). This area falls mostly within Resource Block F and partly in Block A of Ambrose (1973). Within these blocks Kesgrave Sands and Gravels show consistent grading characteristics, with 23 to 25 per cent of the volume of the deposit recorded as gravel grade, i.e. particles greater than 4 mm in diameter. The base of the sand and gravel falls eastwards from just above 20 m OD to 7 m in the Roman River valley and 14 m OD in the Domsey Brook valley. Its thickness is always greater than 7 m; the thinnest sequence (7.3 m) is proved in a borehole at Birch Holt [9157 1881] and the thickest (16.5 m) in another [9450 2079] near Birch Hall.

In the extensive gravel pit at Brake's Farm, Birch, old working faces exposed about 12 m of planar-bedded sand and gravel, with thin lenses of sand up to 10 cm thick and impersistent fine sand and silt bands up to 20 cm thick. The pebble content was similar to that of the Stanway quarries described above.

Where the Kesgrave Sands and Gravels crop out between Birch village [943 200] and the southern limit of the district, there are numerous small pits formerly dug for gravel [9363 1991; 9387 1977; 9276 1905; 9184 1854; 9055 1835].

Yellowish orange fine- to medium-grained sand with gravel was augered on the eastern slope [954 205] of the tributary valley running from the Roman River through Birch village, and an old gravel pit [9562 2098] was noted. Kesgrave Sands and Gravels, 5.5 m thick, resting on London Clay at 27.1 m OD, were proved in a nearby borehole near Conduit Farm [953 203], although to the east no Kesgrave Sands and Gravels appears to be present, and Glacial Sand and Gravel probably rests directly on London Clay.

The sand and gravel mapped out on the northern flank of the London Clay ridge running through Messing-cum-Inworth crops

out from 38 m OD to above 60 m OD. Its base is irregular, a point which is illustrated south of Messing village [897 189]. Here, a borehole [8946 1848] (Haggard, 1972, p.57) proved 16.8 m of pebbly sand resting on London Clay at 43.9 m OD, although only 150 m to the north London Clay crops out at about 54 m OD.

It is likely that Kesgrave Sands and Gravels occur beneath the Terrace Deposits and Boulder Clay in the Blackwater valley and the Domsey Brook, south of the buried channel (see Figure 10); for example, a borehole drilled at Yewtree Farm [8860 1898] proved 8.9 m of sand and gravel (resting on London Clay at 17.3 m OD); this consisted of two distinct units, with 98 per cent sand in the upper 3 m, and between 55 and 75 per cent gravel but only 24 to 44 per cent sand in the lower part.   RAE, CRB

River Colne to the River Stour

North of Halstead [81 30], Pebmarsh [85 33] and Lamarsh [89 36] (see Figure 10), boreholes show that the sequence of Kesgrave Sands and Gravels is similar to that near Sible Hedingham described above. Fine-grained clayey micaceous sands, up to 16 m thick, make up the bulk of the sequence, overlain by comparatively thin sandy gravels.

The gravel pit at Purlshill Plantation [7932 3378] exposed 2 to 3 m of cross-bedded ferruginous sand and gravel overlying 7 to 8 m of fine-grained sand.

The Kilowen sand-pit at Purlshill [800 334] exposed about 10 m of micaceous fine- to medium-grained sands beneath patchy Head. The sands show both cross-bedding and planar-bedding features, and microfaulting locally affects the sequence. Thin pale greenish grey clay partings about 5 mm thick are common; some show evidence of penecontemporaneous brecciation. Certain horizons contain abundant white mica in irregular plates up to 2 mm in diameter.

In the tributary valley to the south of Great Maplestead, the micaceous fine- to medium-grained sands have an extensive outcrop, and gravel beds are thin or absent.

South and east of Halstead, and at Pebmarsh and Lamarsh, boreholes indicate that sandy gravels are the dominant lithology and that the sequence is up to 11 m thick. The gravels are present through much of the area although in two boreholes south of Pebmarsh [8510 3275; 8518 3199] they are absent.

The gravel pit at Ferriers Farm, Bures [895 344], exhibited the following section:

	Thickness
	m
Glacial Sand and Gravel (see p. 32)	9.0
Boulder Clay: sharp planar base; progressively thickens westwards	0.6 to 2.7
Kesgrave Sands and Gravels	
Trough cross-bedded with sets up to 2.5 m thick and shallow angle cross sets, mostly of fine- to coarse-grained sand with some thin greenish grey silt horizons; some ripple drift lamination; pale grey or yellowish white ('Essex White Ballast type'); white vein quartz, patinated flints and bleached quartzite pebbles; orange-brown to rusty brown ferruginous staining patchy; and near the base of the pit there are large blocks of iron-cemented sandstone over 6 m in diameter and 2 m in thickness	up to 10.0

The top part of the Kesgrave Sands and Gravels is clayey and convoluted and is thought to represent the fossilised arctic structure soil described by Rose and Allen (1977) (see Plate 3).

A BGS trial pit [8984 3467] south of Hewitts showed a section in Kesgrave Sands and Gravels:

Plate 3 Boulder Clay overlying Kesgrave Sands and Gravels in the Ferriers Farm pit, Bures. Note the involution structure below the contact

	Thickness m
Top soil	0.3
Gravelly sand, well sorted	0.3
Sand, medium- to coarse-grained, slightly micaceous, orange and pale grey mottled; ferruginous cement present locally in nodular sandstone blocks	1.4
Sand, coarse-grained	seen to 1.2

Mr J. Merritt reported that in the nearby Alphamstone pit [8700 3588] the following section was exposed in 1975:

	Thickness m
Chalky Boulder Clay, pale brown; relatively sharp planar base	7.0
Kesgrave Sands and Gravels	
Sand and gravel, poorly sorted, pale grey to orange; clay in the upper metre or so with involution structures	4.0
Sand and gravel, white, planar bedded and cross-bedded; several channel features; ripple drift cross-lamination locally in finer-grained deposits; partings of pale greenish grey micaceous clayey silt up to 10 mm thick; local brecciation	5.0
London Clay	seen to 0.8

A calcareous concretionary bed below the base of the Boulder Clay was noted at an earlier date.

A BGS trial pit [8998 3354] near Bures proved the following succession of Kesgrave Sands and Gravels:

	Thickness m
Top soil	0.5
Gravel, coarse-grained, sandy, moderately well-sorted; sharp erosive base	0.5
Sand, medium- to coarse-grained, orange brown, iron-stained; local iron pan; local cross-bedding; irregular streaks of clay derived from the bed beneath, below 1.75 m depth; channelled erosive base with a few pebbles	1.5
Clay, silty micaceous, pale grey mottled brown (London Clay)	seen to 0.8

RAE, RDL

In the Wormingford area, sand and gravel between 11 and 13 m thick has been proved in boreholes between the River Stour valley and a line from Janke's Green [904 300] to Wood Hall [937 316]. To the east the sand and gravel thins beneath Boulder Clay. Except for an area at Mount Bures, the Kesgrave Sands and Gravels crop out almost continuously along the Stour valley. The upper part of the deposit is commonly silty and clayey, whereas the lower part, which forms relatively steeply sloping spurs, generally comprises mostly yellowish orange medium-grained sand, the basal part of which is reddish brown in colour and similar to Crag. Boreholes show that the upper clayey part reaches 11 m in thickness beneath the disused Wormingford airfield but thins to less than 5 m near the Stour valley.

In the Great Horkesley area the Kesgrave Sands and Gravels at outcrop are thickest in the Stour valley. Orange and pale yellow medium-grained sand and predominantly rounded gravel, up to 4 m thick, were exposed in the roadside [9723 3302] and a nearby borehole proved 5.6 m.

In the River Colne valley, eastwards from Rose Green [902 281] almost as far as Fordham, the Kesgrave Sands and Gravels crop out extensively. A thickness of 5.5 m, clayey in the bottom half, was proved in one borehole [9026 2834]. Trial pits near Fordham Hall, [9250 2758; 9275 2763] showed the top 2 to 3 m of Kesgrave Sands and Gravels to consist of crudely bedded iron-stained poorly sorted clayey coarse sand and fine to medium gravel with some flint cobbles. East of Fordham, the sand and gravels become attenuated, and they are absent along the valley sides as far as Hillhouse Wood [945 280].   RAE

North of the River Stour

Beneath the Kesgrave Sands and Gravels the bedrock surface falls gently from 50 m OD in the north-west to around 37 m in the east. The thickness of sand and gravel varies between 8 and 17 m except in a borehole [9152 3414], 300 m south of Fysh House Farm, where only 2 m of fine- and medium-grained light brown sand were recorded. At this locality the base of the Boulder Clay falls to around 47 m OD and has cut out most of the Kesgrave Sands and Gravels.

Boreholes indicate the Kesgrave Sands and Gravels generally contain most gravel (up to about 40 per cent) in the upper 2 to 6 m and less than 10 per cent gravel in the lower part which comprises mainly fine to coarse sand. The uppermost beds are commonly clayey, but clayey sand and gravel beds also occur at intervals throughout the lower part of the sequence; thus a borehole [9419 3598], 500 m north of Hullback Farm, proved:

	Thickness m	Depth m
Boulder Clay	3.4	3.4
Glacial Sand and Gravel	3.8	7.2
Boulder Clay	2.3	9.5
Kesgrave Sands and Gravels		
Orange brown sand and gravel	2.2	11.7
Sandy gravel	c.1.8	c.15.5
Yellow-brown sand, medium-grained, silty, and thin clay bands	c.2.0	c.17.5
Sand, silty, becoming coarser grained	c.2.0	c.19.5
Sand, fine- to medium-grained, very silty, brown to greenish brown	c.1.0	c.20.5
Sand, pebbly, dark reddish brown from 21.5 m (? Crag)	c.4.8	c.25.3
London Clay	seen to 1.1	26.4

The lower, more sandy part of the Kesgrave Sands and Gravels was recorded consistently in the area (Hopson, 1981): a maximum of 10.3 m was recorded in a borehole [9651 3627], 300 m south-east of Cock Street.

Orange-brown and yellowish brown, fine- to coarse-grained sands, commonly micaceous, were consistently augered down the more prominent spurs and overlying the London Clay along the Stour valley. Similar deposits were seen in sunken road cuttings at Cuckoo Hill, Bures [911 344], near Pit Field Wood [9273 3405], at High Garth [948 342] and in a road cutting at Harper's Hill which exposed:

	Thickness m
Cross bedded orange-brown coarse sand and gravel	3.0
Dark orange brown sandy gravel	1.0
Pale orange-yellow laminated and cross-bedded fine to coarse micaceous sand with gravelly stringers and some silty fine sand laminae	seen to 3.0
Grey silty clay (London Clay), augered	1.3

GLACIAL SAND AND GRAVEL

These deposits crop out principally east of a line through West Bergholt [960 275], Stanway [940 235] and Birch [94 20], where they generally overlie Kesgrave Sands and Gravels. They form a flat, largely undissected plateau which lies east of the Boulder Clay margin and is probably the final aggradation surface of a glacial outwash plain. An undulating ridge of Glacial Sand and Gravel, about 5 m above the general level of the plateau, trends north–south from Beckingham Hall [933 208] to Brake's Farm [933 198]. The ridge is interpreted as a terminal moraine of the ice-sheet.

Glacial Sand and Gravel has also been distinguished in the Colne valley, near Pool Hill [766 370], where it infills a glacial channel. Smaller tracts, overlying Boulder Clay, occur near Rayne [73 26], between Halstead [81 30] and Wakes Colne [895 285], and north of Bures [907 340]. North and east of the River Stour, Glacial Sand and Gravel crops out from beneath Boulder Clay near Rose Green [93 37] and Hick's Plantation [977 355]. As noted above (p. 17), boreholes have enabled Glacial Sand and Gravel to be recognised below the Boulder Clay elsewhere, in areas where it could not be mapped. This sand and gravel is approximately equivalent to the upper clayey beds of the Kesgrave Sands and Gravels shown in Figure 11.

The Glacial Sand and Gravel is best exposed in the working gravel pits around Stanway, at Warren Lane and White House, where between 2 and 8 m are seen and up to 9.4 m have been recorded in boreholes. Elsewhere, 4.5 m are present at Rayne, 4.6 m near Wakes Colne, and 9 m are exposed in the gravel pit near The Ferriers, Bures [897 337] (see Plate 4). North of the River Stour, boreholes (Hopson, 1981) have proved between 1 and 5 m beneath Boulder Clay.

The Glacial Sand and Gravel is brown, rusty brown or reddish brown (5YR and 10R hues) with pale bluish grey reduction-veining (5B 6/1). This contrasts with the more orange and yellow hues of the Kesgrave Sands and Gravels. Where Glacial Sand and Gravel lies above, or is interbedded with Boulder Clay, the base is generally sharp. Where it directly overlies the Kesgrave Sands and Gravels, the base is gradational and no sharp erosive boundary was seen although the Glacial Sand and Gravels is known to rest on an eroded surface of Kesgrave Sands and Gravel in adjacent districts (Rose, Allen and Hey, 1976; Rose, Sturdy, Allen and Whiteman, 1978). In this district its base is generally drawn in exposures at the lowest clay-rich poorly sorted bed, although it is often impractical to fix a precise horizon. In boreholes, grading curves indicate that there are reworked sediments several metres thick, and the boundary is similarly difficult to place.

Plate 4   Planar-bedded Glacial Sand and Gravel at Ferriers Farm pit, Bures

Lithologically the main feature of the Glacial Sand and Gravel is the relatively high proportion of angular flints, and non-durable rock types such as Chalk, soft siltstone and Jurassic fossils as compared with the Kesgrave Sands and Gravels. The beds are poorly sorted sands and gravels with a clay and silt matrix. They give rise to brown loamy soils with variable amounts of rounded and angular gravel. In the upper part of the Colne valley, on the Chalk bedrock, the deposits contain about 30 per cent of chalk clasts of silt to gravel grade.

In exposures, the Glacial Sand and Gravel varies in lithology vertically and laterally over distances of a metre or less. It is generally weakly bedded and is overall a poorly sorted deposit with as much as 30 per cent by weight of clay and silt. The bulk of the formation consists of fine- to medium-grained sands, some containing randomly oriented fine- to coarse-grade flint pebbles, many of which are subangular and patinated. Other beds are better sorted with weak lamination and pebbly stringers. Beds containing flint cobbles and boulders up to 30 cm diameter also occur, and lenses of pale brown chalky clay (Boulder Clay) have been recorded in the Stanway area; the matrix of these cobble beds is usually a reddish brown or orange brown sandy clay.

In the top part of the formation, in the Stanway area, there are beds of firm red-brown to grey-brown coarse, angular, sandy clay with scattered pebbles mostly of patinated flint; in places faint lamination is seen. These clay beds lithologically resemble decalcified Boulder Clay at outcrop and the basal decalcified Boulder Clay seen in some boreholes (see p. 35). The red hues are similar to those of the upper part of the Kesgrave Sands and Gravels (see p. 22) which are described as a part of a pre-Anglian soil profile (Rose, Allen and Hey, 1976). The red mottling noted in the glacial sediments apparently has a different origin, particularly because in a quarry [950 225] east of Warren Lane, the sediments contained a block of Boulder Clay about 3 m in diameter. It is clear that this reddening took place during the Anglian glaciation or later.   RAE

Details

Western district boundary to the Pods Brook and the River Brain

Between Molehill Green [715 200], Rayne [730 225] and Beddall's End [747 213] there are several tracts of sand and gravel. They are invariably associated with extensive patches of chalk-free (decalcified) sandy clay, but their exact relationships to the chalky Boulder Clay are uncertain. The largest of the tracts, from just north of Bartholomew Green [722 215] to Naylinghurst [736 222], appears to underlie the Boulder Clay. The smaller tracts between Bartholomew Green and Molehill Green, and north of White Court [c.744 218] are thought to overlie the Boulder Clay. The small spreads of sand and gravel [709 203] north-east of Helpestons Manor and at Molehill Green [714 201] give rise to very pebbly soil.

On the side of the Ter valley, two patches of sand and gravel [720 203; 722 201] have been mapped near Milch Hill. The latter patch appears to be dominantly gravelly, and the deposits exceed 1.8 m in thickness. A gravel pit [7218 2003] was formerly opened at this locality. Mottled orange and grey sandy clay was proved locally above sand and gravel near Milch Hill Farm. A much larger outcrop of Glacial Sand and Gravel is present at Bartholomew Green where a gravel pit [721 211] was formerly worked. The surface indications are of a very pebbly deposit at several localities, but yellowish brown clayey silt, yellow sandy clay, and clayey sand and gravel were also noted.

A very small patch of angular flint pebbles has been mapped [702 218] on the Boulder Clay plateau south-west of Graunt Courts. A similar high-level spread [728 201], but less pebbly, is present 800 m ESE of Bartholomew Green. Farther east, to the north of White Court, occurs a spread of orange mottled ferruginous sand and gravel, and grey clayey sand. The sand and gravel is of sufficient thickness and quality to have been worked for domestic use from a small old pit [7432 2181].

It is possible that a spread of sand and gravel [738 221] south-east of Naylinghurst overlies the chalky Boulder Clay, but it cannot readily be separated from a more extensive deposit to the west which underlies the Boulder Clay.

Some 700 m south-east of Hill House Farm there is an outcrop of sand and gravel [755 210] apparently resting on an extensive spread of decalcified till.　　　　　　　　　　　　　　　　　　　　　　CRB

River Blackwater valley

A well [8560 2242] in Coggeshall proved 8.2 m of 'brown sand and olive grey sand' sandwiched between two beds of chalky Boulder Clay (Whitaker and Thresh, 1916, p.123). Other boreholes drilled on the Boulder Clay plateau to the east of Coggeshall proved a similar tripartite sequence, although in these cases the sand and gravel is 3 m or less in thickness.　　　　　　　　　　　　　RAE

River Colne valley

In the upper part of the valley, around Pool Street [766 370], the Glacial Sand and Gravel forms the filling of a glacial channel. Chalky sands and gravels crop out in the upper reaches of the valley and have also been proved in boreholes beneath Alluvium and River Terrace Deposits. At the surface, however, it has not always been possible to separate them from Kesgrave Sands and Gravels.

The railway cutting [7612 3745] south of Great Yeldham formerly exposed over 7.5 m of buff, cross-bedded, fine-grained sand and gravel with lenticular layers of boulder clay (Whitaker and others, 1878, p.47). Comparable lithologies were exposed in a small temporary exposure beside the A604 west of Castle Hedingham [7760 3543].

South of Sible Hedingham, a borehole [7867 3348] proved:

	Thickness m
Second Terrace deposits	
Clayey pebbly sands	1.8
Glacial Sand and Gravel	

	Thickness m
Sandy silts, orange-brown with pockets of angular flints, passing down into sandy gravel	20.1
Boulder Clay	0.3
Upper Chalk	touched

Near Foxborough Hill Farm the following succession was proved in another borehole [7968 3253] sited on the alluvium:

	Thickness m
Alluvium	
Clay and peaty clay	2.4
Sandy gravel	1.4
Glacial Sand and Gravel	
Silt and silty clay, laminated, with gravel bed near the top; chalky sands near base	10.0
Boulder Clay, chalky	2.0
Glacial Sand and Gravel	seen to 0.1

Both the above boreholes proved the continuation of the glacial channel beneath the present-day Colne valley.

Extensive outcrops of Glacial Sand and Gravel deposits are present to the north of Earls Colne where they have been dug from small pits. The deposits are poorly sorted and range from fine-grained sands to coarse gravels. Boreholes prove that the deposits are 3 to 5 m thick, and mapping suggests their base is probably irregular and channelled. The surface of these outcrops ranges between 40 and 62 m OD.

In an outcrop [899 291] north of Chappel, a borehole [8979 2914], sited within 25 m of the chalky Boulder Clay outcrop, proved 4 m of Glacial Sand and Gravel.

Another borehole [8658 2857] in Tey Road, Earls Colne, proved an exceptional thickness of drift deposits:

	Thickness m
Surface level 28.0 m above OD	
River Terrace Deposits	
Sandy gravels with clay beds	7.4
Glacial Sand and Gravel	
Sandy gravels, chalky at base	13.5
London Clay	seen to 0.1

This record is anomalous because there is no other evidence for thick glacial deposits in this part of the valley. The sequence is regarded as the fill of a localised glacial scour.

River Colne to the Roman River

The spread of Glacial Sand and Gravel in the Eight Ash Green area [940 260] gives rise to a silty and sandy soil with angular to subrounded flints. The base of the drift deposits is locally very variable but lies generally at between 34 and 39 m OD. Beyond the eastern limit of the chalky Boulder Clay at Gallows Green [9235 2631] as far as Eight Ash Green, the bulk of the drift is grouped as Glacial Sand and Gravel; only limited areas of Kesgrave Sands and Gravels are present. Boreholes have proved up to a maximum of 7.9 m of Glacial Sand and Gravel [9362 2675]. The deposits consist of brown silty clay and rusty brown clayey sand and gravel.

A BGS trial pit [9406 2621] dug near Fordham Heath exposed:

	Thickness m
Glacial Sand and Gravel	
Fine sand with scattered pebbles of angular to rounded flint and quartzite generally less than 5 cm in diameter, mottled grey to buff; iron-cemented at 0.50 m; structureless; gradational base	0.55

	Thickness
	m
Sandy gravel with a clayey matrix; medium- to coarse-grained sand with gravel as above; orange with grey mottles; clay and gravel content increases with depth	0.65
Lens of orange and grey mottled medium-grained sand with iron-enriched sub-horizontal bands	0.30
Sandy gravel, as above but slightly more clayey	1.30
Reworked London Clay	
Stiff brown clay with grey mottling and a few small flint pebbles	seen 0.55

South and east of the railway line near Eight Ash Green, the Glacial Sand and Gravel is a planar unit; it crops out continuously across a plateau which stretches beyond the eastern boundary of the district. The plateau surface is essentially flat and falls imperceptibly south-eastwards from just above 40 m OD to around 35 m OD. Within this region major gravel operations extract both the Glacial Sand and Gravel and the underlying Kesgrave Sands and Gravels (see p. 25). The Glacial Sand and Gravel in these quarries generally varies in thickness from about 4 to 8 m, although locally only 2 m may be present. Augering proves a reddish or rusty to orange brown sandy and silty clay, and angular and rounded gravel, mostly of flints, is found in the soil. Exposures are numerous, both in small disused gravel pits and cuttings and in the extensive, currently active (1978), sand and gravel quarries in the Warren Lane area of Stanway [950 237 to 949 222]. In these quarries, the Glacial Sand and Gravel is often termed 'hoggin', and is, in places, regarded as overburden.

Old gravel diggings near the railway [9505 2580] about 700 m east of Eight Ash Green exposed variably clayey 'dirty' orange-yellow, silty, fine to coarse sand and angular to rounded mainly flint gravel; the bank to the south of the railway siding showed 5 m of a similar deposit. Nearby borings for the Colchester by-pass in the vicinity of Chitts Hill [962 254] proved between 3 and 4 m of red and brown sand and gravel, and 5 m of orange-brown, silty, fine to coarse sand are present in a road cutting [9625 2547].

In the northern part of the ARC Stanway Bellhouse Farm quarry [952 242], 2.5 to 3 m of Glacial Sand and Gravel were removed as 'hoggin', and the face exposed poorly bedded partially iron-cemented reddish brown coarse sand and fine to coarse, subangular to rounded, mainly flinty gravel. There were also lenses (less than 20 cm thick) of relatively well sorted and laminated medium to coarse sand with only scattered pebbles. The red mottling was comparable in colour to the rubified *sol lessivé* recognised by Rose and Allen (1977) in other gravel pits in this district, to which more detailed reference is made on p. 22. In the main ARC quarry (Plate 1) the working face was about 500 m long and the Glacial Sand and Gravel is rusty brown in colour, contrasting with the pale yellow and orange of the underlying Kesgrave Sands and Gravels. Between 5 and 7 m of Glacial Sand and Gravel are present, and interdigitate with Kesgrave Sands and Gravels in the basal 1 to 4 m, possibly as a result of reworking. A section [9457 2367] showed a typically poorly sorted and roughly bedded sequence of red-brown and orange-brown clayey sands with randomly orientated fine to coarse flints and also some cobble- and boulder-sized flints, up to 20 cm in diameter; pale grey predominantly vertical veining was evident in the upper 2 m of this section and cryoturbation structures occur at intervals along the face. In the same quarry, 5 m of Glacial Sand and Gravel [9507 2337] consist of interbedded 'dirty' and well sorted orange-brown sand overlying a heterogeneous gravelly coarse sand which contains all sizes of gravel from fine-grained patinated flint chips to nodular flint and quartzite cobbles. Pale brown siltstone nodules, less than 5 cm diameter, with concentric banding and commonly with iron-rich cores, occur in the basal metre. A borehole [9404 2386], 600 m west of the working face of the Bellhouse Farm gravel pit, proved ?London Clay beneath 7.6 m of 'Loam' and 'Clay with Gravel', which were classified together as 'Glacial Sand and Gravel' (Ambrose, 1974).

The Francis Aggregates sand and gravel quarry in Warren Lane, Stanway, exposed a similar sequence to that described above. The Glacial Sand and Gravel thins northwards from about 8 m [9452 2255] to some 4 m [9476 2299], and in the currently active (1978) pit [9470 2315] is 6 to 7 m thick. In this new pit the Glacial Sand and Gravel is well-bedded and consists of orange-brown medium to coarse-grained sand with common discontinuous micaceous silt beds up to 20 cm thick; zones of reduction, which probably follow recent root traces, occur down to 3 m. In places [9476 2299], lamination is less distinct and the beds are noticeably more poorly sorted and slightly clayey, with a few seams of grey clay from 1 to 2 cm thick. Cryoturbation structures are common in the upper part of the sequence. Gravel up to cobble size (15 cm diameter maximum) occurs in many beds, and flint constitutes about 90 per cent of the pebbles; weathered quartzite and vein-quartz comprise most of the remaining 10 per cent. Within lenses of finely laminated 'dirty' medium to coarse sand, flint shards and white patinated flint fragments are oriented parallel to the lamination planes. A section in an abandoned face [9452 2255] exposed (in 1976) 4 to 5 m of unbedded mottled orange-brown and grey, sandy, pebbly clay, reddened in places (?decalcified Boulder Clay), with lenses of pale orange cross-bedded sand and gravel and one lens 20 cm thick of gravel, including cobble-sized nodular flints. Beneath were 2 to 3 m of grey-brown chalky sandy clay with pebbles of angular to rounded flint, rounded black Tertiary flint, vein-quartz and some quartzite (chalky Boulder Clay). A discussion of the significance of this occurrence of Boulder Clay within Glacial Sand and Gravel appears on p. 29.

In Hoveringham's Warren Lane gravel pit [951 229], the Glacial Sand and Gravel is 3 to 4 m thick and is characterised by involution structures disturbing its entire thickness. At one locality in the small tributary valley running north from the Roman River [9494 2265], the underlying Kesgrave Sands and Gravels are cut out and the Glacial Sand and Gravel thickens to around 6 m. It consists of dark orange-brown and ferruginous cross-bedded clayey sand and silty sandy gravel. In a disused part of the same Hoveringham quarry [9522 2246] a section, about 4 m high, exposed bedded poorly sorted, orange-brown to orange-red, slightly silty, fine to coarse sand containing variable amounts of gravel, estimated at up to 70 per cent by volume in some beds. The gravel was angular to rounded, and frost shattered in the top metre where it comprised mainly patinated flint; 40 to 50 per cent of the rounded gravel was vein-quartz, the remainder of the rounded gravel and all the angular gravel was flint. One bed 30 cm thick at 2.0 m depth comprised well-sorted medium to coarse sand with subordinate gravel. In the upper 1 m, three continuous seams, 1 to 3 cm thick, of pale bluish grey silt were noted.

Around Stanway Green [957 233] and Shrub End [975 234], the Glacial Sand and Gravel is less clayey than farther west. For example, a small disused gravel pit [9605 2457] exposed 3 m of crudely bedded, cryoturbated, pebbly silty and slightly clayey sand with a thin bed of laminated reddish brown, silty, fine to medium sand containing some fine angular flint gravel. Temporary trench sections in the Shrub End area [973 235] showed up to 3.1 m of orange-brown, poorly sorted, slightly silty, fine to coarse sand with a variable fine to coarse gravel content, mostly of flint. In these sections there was evidence of bedding below 2.0 m depth, although it was affected by cryoturbation.

A disused quarry at Brickwall Farm [966 229] is now a municipal waste-disposal site, but in 1976 the upper part of the main face [9647 2310] exposed bedded, orange-brown, partially iron-cemented, silty, fine to coarse sands, with variable amounts of gravel of which 90 to 95 per cent was less than 2 cm in diameter. An ice-wedge cast with a zone of disturbance 1 m wide extended from the surface down to 4.5 m.

Glacial Sand and Gravel about 3 m thick, displaying similar

lithologies to the pit at Brickwall Farm, was noted in temporary sections [9798 2265; 9787 2247].   RAE

Roman River to the southern district boundary

West of the Roman River, small tracts of Glacial Sand and Gravel [932 238; 932 229; 936 220] are characterised by gravelly soil containing mostly angular to subrounded patinated flints. The deposit is probably continuous beneath the Boulder Clay in places, as shown by a borehole [9349 2255] which proved 'very clayey fine sand' and 'silty sandy clay' totalling 2.7 m, beneath chalky Boulder Clay and overlying Kesgrave Sands and Gravels.

Larger areas of Glacial Sand and Gravel have been mapped on the plateau south of Heckfordbridge [946 217] as far as Birch village [943 200], where orange, red-brown and grey, very sandy clay was augered beneath a flinty soil. A number of low ridges and mounds rising some 4 to 5 m above the general plateau level form relatively prominent features in this region. They are characterised by very gravelly soil with mainly angular to rounded flints. A BGS trial pit on the flank of one ridge proved:

	Thickness m
Glacial Sand and Gravel	
Sandy gravel, slightly clayey in patches; mainly unbedded; mottled orange-yellow to orange-brown with variable iron-staining; the gravel is of fine to coarse (up to 8 cm diameter), mainly subrounded to rounded flints with subordinate quartzites and vein-quartz pebbles; the sand is mainly coarse to very coarse and occur in lenses near the base up to 10 cm thick	2.50
Sand, silty and slightly clayey, bedded, medium- to coarse-grained with scattered fine to coarse angular flints; a few pale orange-yellow silt beds up to 10 cm thick	seen to 1.50

The ARC gravel pit at Brakes Farm, Birch [927 200] exposed 4 m of clayey Glacial Sand and Gravel near the entrance of the pit from the main road:

	Thickness m
Red-brown poorly sorted variably clayey sand and gravel; crudely bedded with gravel-free medium sand lenses; convoluted base in places	1.5 to 4.5
Laminated fine to medium silty sand with flint and quartzite cobbles at base (cut out by bed above in places)	0 to 2.0

In the same quarry the following strata were exposed in the upper part of the Glacial Sand and Gravel:

	Thickness m
Reddish brown sandy clay, laminated in places (?decalcified Boulder Clay); slightly cryoturbated base with minor undulations	2.0
Dark purplish grey highly brecciated clay with scattered bleached flints; interbedded with orange-brown clayey and silty sand at base	seen to 0.3

RAE

River Colne to the River Stour

On the west side of the Stour valley, between Henny Street and Lamarsh, there are three outcrops of Glacial Sand and Gravel. A BGS trial pit near Hill Farm on the central outcrop [8785 3641] showed the sequence and lithology in detail:

	Thickness m	Depth m
Top soil, pebbly	0.20	0.20
Gravel in brown sandy clay to clayey sand matrix, unbedded; angular and rounded flints and quartzites along with some flint cobbles near sharp base	1.00	1.20
Clay, orange and grey mottled	0.02	1.22
Gravel, sandy and clayey, brown, with angular and rounded flints up to 2 cm diameter	0.13	1.35
Sand, fine-grained, orange-brown; poorly sorted gravels and clay pellets in basal beds	0.35	1.70
Sand, fine- to coarse-grained, well-sorted, with dominantly rounded gravels (Kesgrave Sands and Gravels)	1.70	3.40

The gravel beds show evidence of reworking, and more clayey beds at 1.20 and 1.70 m may be the clayey bases of solifluction deposits.

An outcrop of gravel at Ferrier's Farm Pit [895 344] provided the best exposure of sandy Glacial Sand and Gravel in the district (Plate 4). Up to 9 m were exposed, consisting of planar bedded, orange-brown, sandy gravel containing subangular flint and subordinate vein-quartz and quartzite clasts. The deposit is moderately well-sorted near the base where there is some cross-stratification. Poorly sorted ochreous clayey sand beds up to 10 cm thick occur throughout. In the uppermost 2 to 3 m the deposit is cryoturbated, poorly sorted and clayey. In the eastern part of the pit it oversteps the underlying Boulder Clay to rest directly on Kesgrave Sands and Gravels.

To the east of Halstead, a borehole [8228 2932] proved 5.5 m of gravel resting on chalky Boulder Clay. The surface lithology is fine-grained sand.   RAE, RDL

From The Chantry [9724 3246] to Great Horkesley [977 305] the soils are characteristically gravelly, and deposits varying from silty sand and gravel to sandy clay were augered.

A roadside ditch [9725 3299 to 9729 3284] on Harkers Hill, Nayland, exposed orange-brown silty fine sand and clayey fine to medium sand, overlain by orange medium sand with mostly rounded gravel comprising mainly flint with some quartz and quartzite pebbles. A nearby borehole [9736 3262] proved 5.0 m of Glacial Sand and Gravel comprising beds of orange-brown to pale yellow-brown very clayey sand, with only scattered flints and subordinate more gravelly clayey sand with flints, vein-quartz and quartzite pebbles. Field drains near The Chantry [9725 3220] exposed variably clayey, orange-brown and red-brown, medium to coarse sand with mostly rounded gravel; in places flint cobbles and angular flints were found.

The Glacial Sand and Gravel consists mainly of sand in the area east of the Roman Road into Great Horkesley. Temporary excavations [9782 3157] showed 1.5 m of brown and grey mottled very sandy clay and clayey fine to coarse sand with layers of clay-free sand and gravel 20 cm thick; the sandy beds contained small hard chalk pellets up to 1 mm diameter. An adjacent borehole penetrated 3.2 m of very clayey sand overlying Kesgrave Sands and Gravels.

The tract of Glacial Sand and Gravel west of Great Horkesley is clayey, as shown by trial pits dug near Breewood Hall. Sandy clay and clayey sand, 4 m thick, overlie sandy gravel [9719 3082] referable to the Kesgrave Sands and Gravels. A little to the north, a further trial pit proved orange-brown and grey-blue mottled, very sandy clay and clayey fine to medium sand with a variable gravel content; the gravel was mostly medium and coarse flint much of which was patinated.

From south of Great Horkesley to Pitchbury Ramparts [966 290], the Glacial Sand and Gravel has a variable lithology although augering proved that medium to coarse sand predominates. Its base is not clearly defined but, in places [9780 2878; 9635 2922], it rests directly on London Clay. Two boreholes nearby proved 5.2 m and

7.9 m of silty clays (Ambrose, 1974, p.38) which are now included in the Glacial Sand and Gravel.

Around West Bergholt [960 275] a clayey deposit, similar to a decalcified Boulder Clay, has been augered, and orange-brown flinty silty and sandy clay was also noted [9560 2791].

Boreholes and a BGS trial pit suggest that Glacial Sand and Gravel fills a local hollow in the London Clay surface beneath the Head mapped on the valley side [955 273]; it was proved in one borehole which passed through 9.8 m of brown silty clay and gravel (see Ambrose, 1974, p.35). A BGS trial pit near the Old Rectory [9553 2732] exposed 3.25 m of poorly bedded, orange, fine to coarse well-rounded gravel in a coarse sandy matrix which was locally clayey; some cobbles, up to 20 cm in diameter, of well-rounded vein-quartz and subangular flints were also noted.

A maximum of 8 m of Glacial Sand and Gravel was recorded in boreholes north of Janke's Green [905 298], but generally less than 3 m are present. RAE

North of the River Stour

Glacial Sand and Gravel has been mapped in the north-eastern part of the area, and boreholes have proved it beneath the Boulder Clay. It crops out along the extreme eastern part of the Stour valley between Harpers Farm [9640 3496] and the district limit, and in a tributary valley running south from around Dorking Tye [920 367] and Mill Farm [936 369]. The gravels vary in thickness, and are up to 8.5 m thick [9022 3701]. At outcrop they are characterised by a variably clayey reddish brown sandy soil with numerous angular flints. The deposit has in places undergone mass movement downslope and obscures the underlying Kesgrave Sands and Gravels [9790 3535]. Clayey lenses occur within the gravels, and these have in a few places been mapped as Boulder Clay, for instance east of Honey Tye Farms [966 367] and about 500 m southeast of Cock Street, [9660 3615].

A BGS trial pit [9786 3694], about 400 m south of Stoke Priory, proved the following sequence of Glacial Sand and Gravel:

	Thickness m
Reddish brown unbedded gravelly and clayey fine to coarse sand; the gravel is subangular to rounded, fine to coarse and mainly of flints some of which are nodular; the flints are generally brittle and easily shatter into angular fragments; lenses of medium and coarse sand are present, up to 8 cm thick and 30 cm long; the clay content is variable, and many of the flints have a thin coating of pale grey-blue clay; patches of stiff sandy clay occur throughout (?decalcified Boulder Clay)	seen 3.40

A further pit [9671 3510] near Harper's Hill, Nayland, proved:

	Thickness m
Glacial Sand and Gravel	
Sand and gravel with clay lenses; many large nodular and subangular flints between 10 and 15 cm diameter	1.15
Sandy gravel, orange-brown; some horizons dominantly sand; the gravel is mainly fine- to medium-grained flints with some quartz and quartzite	0.55
Sand, medium- to coarse-grained, orange, with scattered flints (<1 cm diameter), iron-stained; vague bedding traces	0.45
Gravel, sandy, orange-brown with red mottling; increasing clay content with depth associated with grey mottling; medium grained sand lenses up to 10 cm thick	seen 1.25

RAE

BOULDER CLAY

Boulder Clay is the most widespread drift deposit in this district covering about 60 per cent of the area (Figure 12). Over most of its outcrop it forms a rather monotonous featureless plateau. At three localities (south of Sible Hedingham [782 241]; in the Pant–Blackwater valley between Shalford [724 292] and Stisted [799 246], and in Pods Brook near Clapbridge Farm [742 227] it forms outliers at topographically lower levels along valley sides, where it probably represents the eroded remnants of a Boulder Clay infilling to the valleys. The Boulder Clay is generally sheet-like in form but, as a result of dissection by streams, the outcrop has a very irregular margin on the valley sides; only at one locality, in the River Pant valley north-west of Clapbridge Farm, has downcutting not penetrated the Boulder Clay. Boulder Clay also locally occurs beneath Glacial Sand and Gravel between Bartholomew Green [722 215] and Molehill Green [713 198], to the south of White Court [742 210], south of Colne Park [872 302], at Hill Farm [881 381], and just east of Speck's Farm [892 345]. Boreholes have shown that Boulder Clay also occupies buried glacial channels in the River Colne valley north of Brook Street Farm [810 319], in the Roman River valley between Great Tey [892 257] and Stanway [940 242], and in the Blackwater valley at Kelvedon [861 188] (see Figure 12). At these three localities the Boulder Clay is overlain by river terrace gravels and alluvium, and in the latter two valleys it is also overlain by Hoxnian interglacial silts and clays (see p. 40). There are two contributory factors to the occurrence of Boulder Clay in the valleys: the first is primary deposition of the Boulder Clay in pre-existing valleys which have since been partly reexcavated by the modern rivers to leave 'buried channels' infilled with Boulder Clay; and the second is creep and solifluction along with cambering on the valley sides, which may have resulted in mass downslope movement of the Boulder Clay. Boulder Clay is absent east of a line running approximately from Nayland through Little Horkesley [961 320], West Bergholt [962 278], Stanway [940 242], and Birch [944 200]. This margin is regarded as the easternmost limit of the Anglian ice-sheet.

The base of the Boulder Clay maps out as a sharp planar boundary which, in places, is followed by a slight feature that is locally more pronounced on outliers in valleys. Where seen during the recent survey the base was generally sharp on underlying sands and gravels; there was no exposure of Boulder Clay resting directly on London Clay although this superposition has been proved by augering and in boreholes in several areas (see Figure 11). The general trend of its base has been constructed across the district using data from mapping and from several hundred borings. The resultant contour plot, shown in Figure 12, illustrates an overall fall both north-west and south-east from a broad sub-Boulder Clay ridge, at 70 and 85 m OD, running through Castle Hedingham [785 355] and Wethersfield [712 312]. To the north-west of this ridge it falls relatively quickly to 40 m OD, although this represents only a slope of less than 0.5°. To the south and east there is an even more gradual fall to around 30 m OD at the eastern margin of the Boulder Clay outcrop; this gradient is interrupted by a step-down to about 20 m between Cressing [792 208], Coggeshall [853 230] and Great

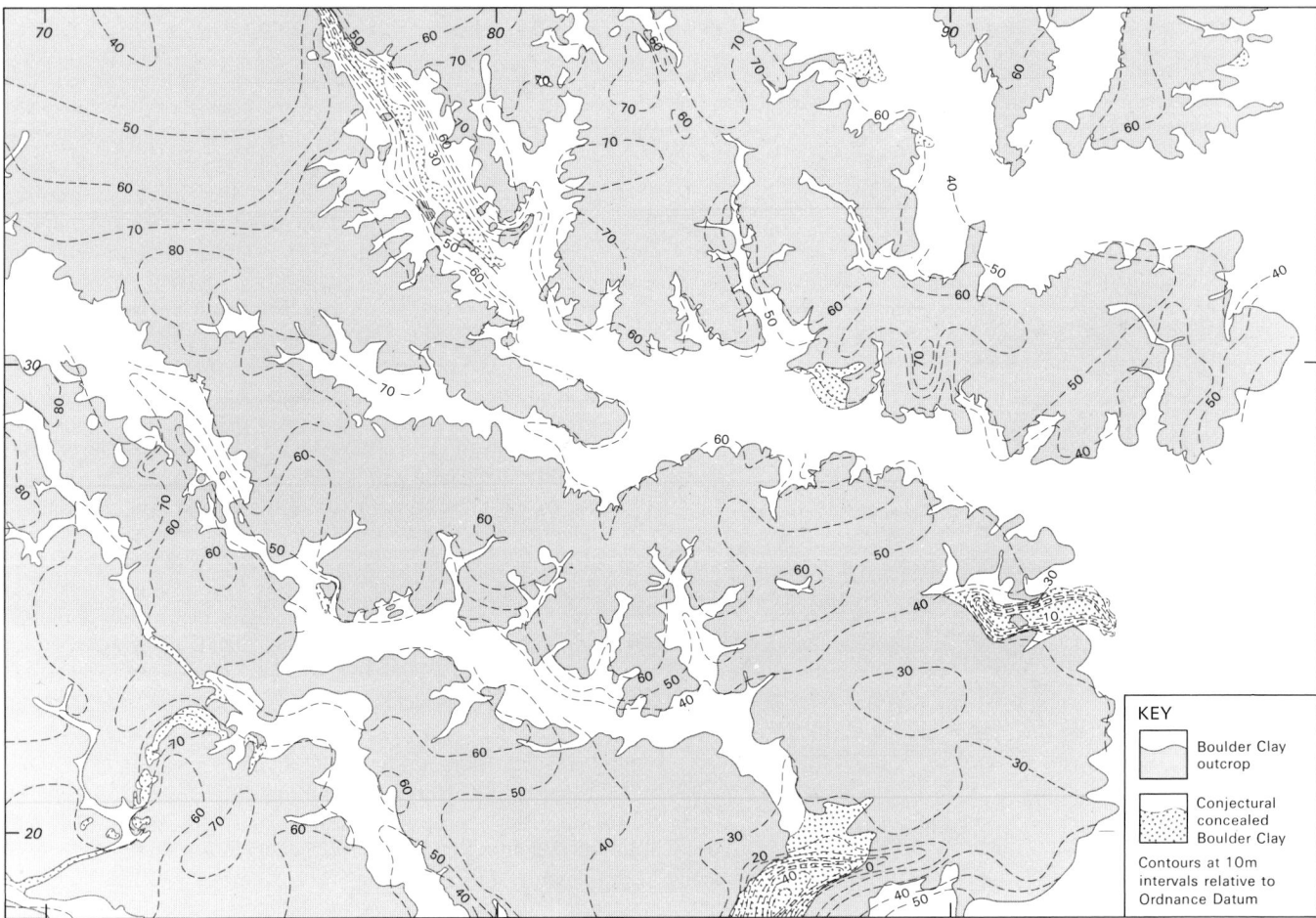

Figure 12 Contours on the base of the Boulder Clay

Tey [892 258], although even here the slope is only about 0.5°. Superimposed upon this gentle regional gradient are subsidiary falls in the base of Boulder Clay into the valleys , where the base may be up to 15 m lower than beneath the interfluves.

The base of the Boulder Clay falls much more rapidly into the glacial channels, with gradients as steep as 18°. The maximum depth of these channels has not been proved but in the Colne valley [7968 3253] the base of the Boulder Clay falls to lower than 22 m OD. At Marks Tey, one borehole [9066 2473] terminated in Boulder Clay at 11 m below OD, and a further one [875 190] at Kelvedon proved it to 33.2 m below OD, falling further to 51 m below OD to the south of this district. Within these channels the Boulder Clay is locally interbedded with laminated clays and silts of probable subglacial origin.

The gentle rise in the base of Boulder Clay in the south of the district (see Figure 12), is caused by a London Clay ridge at Messing [899 189] which continues in a south-westward direction towards Tiptree across the adjacent Chelmsford district, and is thought to have been an effective barrier to the southward movement of the Anglian ice-sheet.

Although the original thickness of Boulder Clay is not known because of post-Anglian erosion, a general statement regarding the considerable variation in thickness is given here and shown diagrammatically in Figure 13. In general, erosion has resulted in the greatest thicknesses being pre-served along the interfluves. The maximum proved thickness of 28 m is at Bradfield [726 365] in the north-west but, because this figure was recorded from a borehole sited in a shallow valley on the Pant–Colne interfluve, a maximum figure of 50 m can be expected beneath the Boulder Clay plateau where it is highest. The thickness across the plateau between Wethersfield and Toppesfield and south towards Liston Hall Farm [767 313] is everywhere greater than 15 m and across large areas of the interfluve between Great Saling [702 255], Panfield [739 253] and High Garrett [777 272], and south of Little Maplestead [824 340], it exceeds 18 m. In the more eastern parts of the district the deposit is generally less than 12 m thick although more than 18 m are present near Fordham Place [939 292], Pond Farm [954 295], and in the Domsey Brook valley. The thickness of Boulder Clay in the buried channels cannot be determined precisely either because it is interbedded with Glacial Sand and Gravel and Glacial Silt or because its base has not been penetrated. It is, however, likely that the lowest drift deposit in these buried channels is Boulder Clay, and this assumption is borne out by the sparse borehole data available in the Kelvedon buried channel where a maximum of 49.1 m of Boulder Clay has been proved overlying the Lower London Tertiaries.

The Boulder Clay is a pale brown to buff (10YR hues), very chalky, sandy clay over most of its outcrop, although near the surface it is decalcified and leached to a mottled dark reddish or orange-brown (10R to 5YR hues) and pale grey (N6) colour in places. The thickness of this weathered layer varies considerably, but over most of the outcrop rarely exceeds 1.5 m. Several interactive factors have a bearing on the depth of weathering; they are the degree of slope of the ground surface (and hence the rate of run off), the position of the water table, the degree of fissuring, and the history of vegetation cover.

A comprehensive suite of erratics has been noted during the survey. The most abundant clasts are of chalk, and these range from about 5 cm in diameter down to small pellets and a silt-grade chalk 'flour'. Perrin and others (1973) estimated that chalk comprises 56 to 84 per cent of the erratic content of the Boulder Clay. The remaining erratics include iron-cemented sands and boxstones, subangular to rounded and nodular flints, well-rounded black flints derived from Tertiary beds, Tertiary 'sarsens' (silica-cemented sandstones) which reach up to 2 m in diameter and are the largest erratics found in this district, rounded vein-quartzes, Cretaceous sandstones, cream and yellowish brown Jurassic limestones, Kimmeridge Clay clasts and calcareous nodules, Bunter quartzites, Carboniferous limestones, and igneous and metamorphic rocks. Jurassic and Cretaceous fossil fragments, particularly belemnite fragments, *Inoceramus* and echinoid fragments derived from the Chalk have also been found. No quantitative work has been undertaken during this survey but Perrin and others (1973) noted that, of the erratics other than chalk, flint clasts and 'others' constitute about equal proportions, and igneous and metamorphic rocks make up about 0.1 per cent of the 'others' (see also Boswell, 1929; Baden-Powell, 1948; and Clayton, 1957).

The upper pale brown weathered part of the Boulder Clay grades down into grey (N3) coloured Boulder Clay over a mottled transition zone which may reach 2 m in thickness and which occurs at depths ranging from 3 to 11 m below the surface. The weathered layers in general appear to contain a greater proportion of chalk clasts compared with the grey Boulder Clay beneath, although the carbonate content does not vary significantly between the two (Perrin and others, 1973). This apparent chalk enrichment in the upper layers may, therefore, be a function of the comminution of the chalk clasts.

In general, thick sequences (greater than 10 to 15 m) of Boulder Clay are grey in colour. This lithology is a stiff sandy clay with an erratic suite the same as that described above from the surface layers. Thin, interbedded, gravelly bands, orange to rusty brown in colour, occur locally, and there are slight colour variations from dark grey (N3) and dark brownish grey (5YR 3/1) to olive grey (5Y 4/1 to 5/1) which are presumably governed by the rocks over which the ice-sheet has moved. Most obvious are the darker grey bands which contain a higher proportion of dark grey fissile Kimmeridge Clay clasts up to 5 cm in diameter. The proportion of chalk and other erratics varies vertically and laterally although there is no consistent pattern, and both an increase and decrease of chalk with depth has been observed in different boreholes.

The basal part of the Boulder Clay, generally a bed 0.1 to 0.5 m thick, is commonly dark brown (5YR 3/1 to 10YR 3/2,) and is either chalk-free or contains only scattered pellets. The proportion of clay in the matrix is lower than in the succeeding beds, whilst the amount of sand and gravel clasts is proportionally increased, probably reflecting incorporation of underlying sand and gravel. Crude lamination is present in places and, along with the horizontal orientation of the pebble long-axes, suggests deposition in a viscous or fluid state, possibly as a flow-till. The reason for the brown colour is imperfectly understood, but it is clearly related to oxidation of the clay matrix of the Boulder Clay. To understand this phenomenon the clay mineralogy of the matrix needs further study, although Perrin and others (1973) have recorded mica, kaolinite, variable smectite and locally chlorite—a suite not inconsistent with derivation from Jurassic clays, Gault and Chalk. On the other hand, surface weathering of the grey Oxford Clay, Kimmeridge Clay and Gault oxidises them to orange and grey colours, not to dark brown; it is thus improbable that subaerial weathering is responsible for the colour. One explanation is that partial oxidation of these basal beds occurred during the post-glacial period when the underlying sand and gravel was saturated, so that the piezometric level was above the base of the Boulder Clay and thus a high pore-water pressure existed in the basal beds of the Boulder Clay. The reduced proportion, or lack of chalk in the basal beds of the Boulder Clay may also be a primary depositional feature; because of the increased solubility of calcium carbonate in cold water, chalk debris in the base of the ice-sheet may have removed prior to deposition of the basal till.

The Boulder Clay is overconsolidated and is likely, therefore, to have a low compressibility making it generally a suitable foundation for large structures. The state of consolidation of till is determined by it stress history, which is governed by its depositional and post-depositional geological history (Boulton and Paul, 1976). A subglacial lodgement till, such as the Boulder Clay of this district, has a stress history reflecting the growth of the ice-sheet and the subglacial water pressure, both of which have a bearing on the drainage conditions beneath the ice-sheet. The fluctuations in stress beneath the ice-sheet are likely to have been considerable and may not have been necessarily related to the thickness of ice; thus the degree of overconsolidation of the Boulder Clay cannot be relied upon to reflect the thickness of the Anglian ice-sheet.   RAE

Details

Western district boundary to the Pods Brook and the River Brain

Within this area there is an overall fall in the level of the Boulder Clay base from 80 m OD to 60 m OD (Figure 12). This general picture is locally modified by a subsidiary fall into the major valleys and by a sharp rise associated with a complicated sequence of drift deposits between Rayne [730 228] and Thistley Green [703 195]. Near Young's End [740 195] there is a rise to 70 m OD, before the base of the Boulder Clay continues its southwards fall across the Chelmsford district (Bristow, 1985, fig. 17). Although in general the base of the till is planar, there are marked variations in its thickness, and the overall thinning from the plateau crest to the valley sides is irregular. The areas of thick Boulder Clay usually coincide with areas in which the Kesgrave Sands and Gravels are thin or absent (Figures 10, 13). The maximum recorded thickness of the till in boreholes west of the River Brain is in excess of 18 m [7074 2468]; [7200 2251]; [7242 2315]; [7157 1945].

Over most of the area the Boulder Clay is typically very chalky to within a few centimetres of the surface. Locally, however, there are extensive patches of chalk-free mottled orange and grey sandy clays which are interpreted as decalcified Boulder Clay. The exact relationship of this chalk-free material to the adjacent deposits is unknown, but it is invariably associated with deposits of Glacial Sand and Gravel which extend from Molehill Green [715 200] through Rayne [730 225] to Beddall's End [747 213] and, farther east, towards Lodge Farm [759 220]. In one borehole [7438 2183], some 2.4 m of brown and grey mottled clay overlie typical chalky Boulder Clay (Clark and Ambrose, 1975).

No Boulder Clay has been mapped beneath sand and gravel in the Braintree area, although the configuration of spreads of Glacial Sand and Gravel west of Hill House Farm [c 745 215] suggests that a lower Boulder Clay may crop in this vicinity.   CRB

The Pods Brook to the River Pant

The chalky Boulder Clay is more than 18 m thick beneath the interfluve near Panfield. To the north around Iron Bridge Farm, Shalford (see Figure 13) the thickest deposits (up to 14 m) are probably associated with a channel-fill in which the Boulder Clay apparently has a markedly irregular base. Across most of the area Boulder Clay lies on sand and gravel, but the following examples prove that Boulder Clay occurs locally both above and below sand and gravel.

One borehole [7208 2634], south of Jasper's Green, proved the following anomalous sequence (Clarke and Ambrose, 1975):

	Thickness m
Boulder Clay, chalky	3.4
Glacial Sand and Gravel	6.1
Boulder Clay, chalky	2.4
London Clay	seen to 0.3

No evidence of this lower Boulder Clay is known at outcrop in this area, although a comparable deposit has been recorded to the east and south-east (see p. 37 and below).

Another borehole [7537 2471] at Panfield Lane, Braintree, proved:

	Thickness m
Boulder Clay, chalky	4.9
Glacial Sand and Gravel	4.0
Boulder Clay, chalky	5.5
London Clay	seen to 0.9

Again no evidence of the lower Boulder Clay is found at outcrop, but this section makes an interesting comparison with the nearby deposits around Doreward's Hall and Strait's Mill (see p. 24) where the glacial sequences are anomalous.   RDL

The River Pant to the River Colne

The thickest Boulder Clay sequence in the district is present around Toppesfield where 29.6 m were proved in a well [7386 3743]. Thick deposits of Boulder Clay are also present to the south-east, near

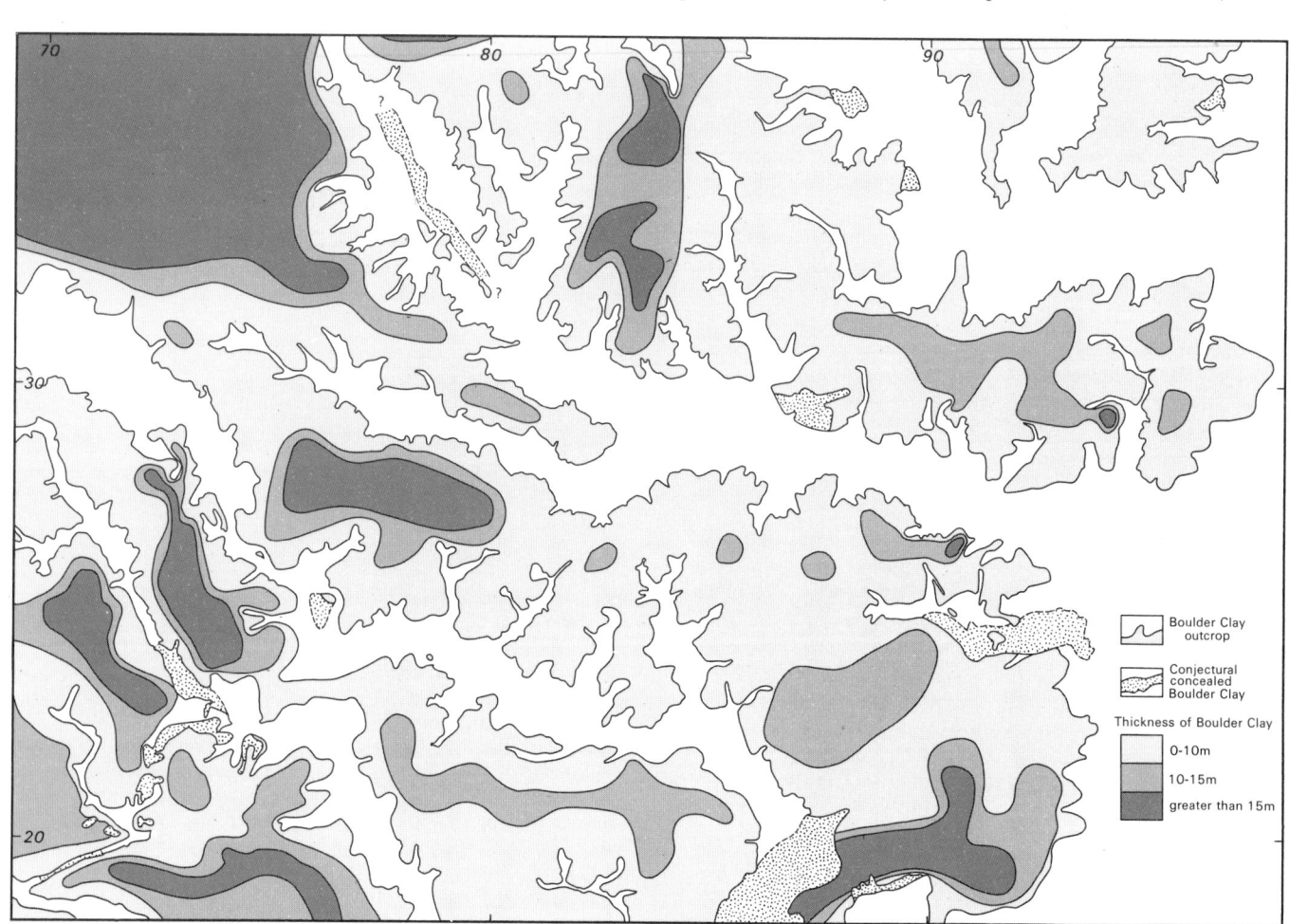

Figure 13   Generalised Boulder Clay isopachytes

High Garrett (see Figure 13) where boreholes have proved over 18 m.

Trial boreholes near Upper Wright's Farm, north of Blackmore End, have proved intercalations of laminated clays and sandy silts within the Boulder Clay, notably in three boreholes [7392 3250; 7455 3194; 7549 3213].

The Beazley End gravel pit [738 289], an extensive gravel working has been extensively backfilled and regraded; it showed the following succession in 1972:

	Thickness m
Brown and grey clay with abundant flints, passing laterally into	1.3 to 1.5
Dark grey sandy chalky clay (not present in southern part of face)	c. 2.1
Grey to yellow brown chalky sandy clay; sharp irregular base	c. 1.5 to 2.1
Kesgrave Sands and Gravels	seen to 1.8

At Foxborough Hill gravel pit [793 320], 1.0 to 1.5 m of grey chalky Boulder Clay is overlain by 0.5 to 1.0 m of distinctive paler and more silty Boulder Clay, rich in chalk flour, visible at the top of the old working face. Immediately west of Purlshill [793 338] a small disused sand pit exposed about 1 m of brown decalcified Boulder Clay with a sharp base, underlain immediately by a calcrete horizon up to 5 cm thick.

A well [8562 2284] in Coggeshall proved an anomalous succession of drift-deposits as follows:

	Thickness m
Chalky Boulder Clay	2.1
Glacial Sand and Gravel	8.2
Chalky Boulder Clay	11.3
London Clay	—

Two boreholes [8795 2284; 8794 2115] to the east of Coggeshall have also proved a lower Boulder Clay; the latter borehole penetrated sand and gravel between the lower unit and the London Clay. It is probable that the lower till is extremely local in occurrence in view of the other borehole evidence available (Booth and Merritt, 1982).

The base of the Boulder Clay is markedly channel-like in form in the Sible Hedingham area, and the axis of the channel corresponds with that of the present-day valley (see Figure 12).

The drift deposits downstream from Halstead generally show a simple layer-cake stratigraphy: Boulder Clay overlies Kesgrave Sands and Gravels which rest on London Clay; near Earl's Colne however the Boulder Clay rests directly on bedrock.   RDL

The River Blackwater to the Roman River

In this area the base of the Boulder Clay falls from 40 m OD to just below 30 m OD at its eastern limit. Superimposed upon this general trend is a steep fall into the glacial channel at Kelvedon, where its base is 50 m below OD under the terrace deposits of the Blackwater, and a fall into the basin at Marks Tey to below OD (see Figure 12). In the Kelvedon area it is difficult to interpret logs penetrating the glacial channel deposits, for these contain both glacial lacustrine sediments and chalky Boulder Clay: for convenience the entire sequence has been grouped as Boulder Clay. The Boulder Clay varies considerably in thickness from over 18 m in the vicinity of Domsey Brook to around 5 m near its contact with the Glacial Sand and Gravel in the east; the relationship between the chalky Boulder Clay and the Glacial Sand and Gravel is discussed more fully on p. 32.

Near Copford Green [928 226] the Boulder Clay is decalcified in places, and rests directly on London Clay, as shown for example by a borehole [9269 2338] which proved 4.9 m of Boulder Clay,

described as 'brown silty clay', on London Clay (Ambrose, 1974, p.46). In general in this neighbourhood, the Boulder Clay is oxidised to an orange-brown colour to a depth of about 5 m, below which it assumes the characteristic grey colour.

Across the plateau, in a tract from Coggeshall to Copford and to its southern limit, the Boulder Clay is typically very chalky either at the surface or beneath a decalcified layer which is normally only about 1.5 m thick; old marl diggings are common in many of the fields, particularly near to farm buildings. Thin gravelly bands occur within the Boulder Clay as in a borehole [9050 2131] north-east of Badcock's Farm, where a gravelly clay 0.3 m thick was encountered from 9.1 to 9.4 m. The basal 1.2 m of chalky Boulder Clay in this borehole is brown in colour and rests on London Clay. Similarly, three boreholes [8796 2284; 8795 2114; 8889 2331] proved respectively 3.0 m, 2.0 m and 0.9 m of gravel within the Boulder Clay, although such thin gravel bands have not been mapped at surface; the first mentioned boring proved 24.0 m of drift without reaching bedrock (see Booth and Merritt, 1982, p.86 for details).

Near Kelvedon, chalky Boulder Clay lies beneath terrace deposits. A small gravel pit at Little London [8719 1992] near Feering had chalky Boulder Clay exposed in its bottom, and the A12 Kelvedon By-Pass cutting [877 196], to the east of Kelvedon, exposed ferruginous sand and gravel overlying chalky Boulder Clay, which was also penetrated to 18.9 m without proving bedrock in a nearby borehole [8776 1956]. Neither of two nearby boreholes [8639 1867; 8710 1926] entered bedrock, but penetrated terrace deposits and stopped in chalky Boulder Clay at 7.3 and 5.8 m above OD respectively (see Haggard, 1972).

A well [8046 1915] at the now disused Fullers Brewery in Kelvedon proved 48.7 m of Boulder Clay beneath gravels, resting on ?London Clay at 33.2 m below OD (Whitaker and Thresh, 1916, p.202). Another well [8758 1902], south of Threshelfords Farm, penetrated the following drift deposits overlying the Lower London Tertiary deposits at 27.4 m below OD:

	Thickness m
Top soil	0.3
Loamy clay	0.6
Gravel	0.9
Clay and chalk stones	49.1

Only 100 m to the south a borehole [8764 1894] proved 10.7 m of chalky Boulder Clay, whose base was at + 14.3 m overlying Glacial Sand and Gravel. A well [8721 1966] at Gore Pit, Feering, proved the base of the Boulder Clay at about 18 m above OD, overlying Glacial Sand and Gravel (Whitaker and Thresh, 1916, p.155).

Beneath the chalky Boulder Clay spread extending from Prested Hall [883 197] to Messing [895 191], there is more than 18.3 m of till in two boreholes [8831 1989; 8956 1940] (Haggard, 1972) which fills the shallowing eastern extension of the Kelvedon buried channel; nearby boreholes (Ambrose, 1974) have shown the Boulder Clay to rest directly on London Clay at elevations of 18.0 and 22.9 m above OD respectively (see Figure 12).

South of the Domsey Brook the Boulder Clay surface rises to above 45 m OD at Inworth Hall [8782 1819] near where a borehole proved 2.7 m of brown chalky Boulder Clay resting on sand and gravel at 41.8 m above OD.

Extensive tracts of Fifth Terrace overlie the chalky Boulder Clay around Messing, although in places the gravelly terrace deposits are thin; chalky clay was augered beneath them [8862 1919; 8865 1909]. A borehole [8860 1898] proved only 0.9 m of gravelly soil on 7.6 m of Boulder Clay and 8.9 m of sand and gravel; the base of the till here is at 26.2 m above OD.

From Harborough Hall [9005 1867] to near Brakes Farm [9300 1981], the limit of outcrop is marked by numerous shallow pits or hollows [e.g. 9042 1855; 9096 1839; 9138 1846; 9254 1930; 9284

1995], formerly excavated for marl as top dressing on the fields to the south which are underlain by London Clay.

Boulder Clay, both chalky and decalcified, is interbedded with Glacial Sand and Gravel at a number of localities; a borehole [9438 2309] proved 'brown clay with flint, becoming chalky between 3.7 and 5.2 m', within a deposit described by Dalton (1880) as 'Loam' but which is now interpreted as Glacial Sand and Gravel (see Ambrose, 1974, p.52).

In the Stanway Warren Lane gravel pit [9452 2255], owned by Francis Aggregates a section in 1976 showed Glacial Sand and Gravel (described on p. 31) containing irregular beds of orange-brown and grey mottled, flinty sandy clay, which are interpreted as decalcified Boulder Clay, and also an irregularly shaped pod of grey-brown chalky Boulder Clay apparently cutting across the bedding.   RAE

The River Colne to the River Stour

In this area the Boulder Clay is generally no more than 18 m thick. Boreholes indicate that the typical lithologies are consistently present, and only one hole has proved an anomalous sequence.

There is an outlier of Boulder Clay below the level of the plateau in the River Colne valley in the vicinity of Dyne's Hall [804 327], with smaller areas nearby [7995 3290; 7965 3330; 7970 3355] where old marl diggings were noted during the present survey. A borehole [8029 3270], near Dyne's Hall, proved:

	Thickness m
Clay, sandy and flinty, brown and grey mottled (decalcified Boulder Clay)	3.6
Clay, sandy clay and silt, orange-brown, interbedded and locally finely laminated	4.6
Clay, sandy, with mainly chalk clasts, orange-brown in the top 0.4 m, then grey (Boulder Clay)	9.1
London Clay	seen to 1.0

This composite sequence is probably part of the filling of the nearby glacial channel in the Colne valley (see also pp. 30 and 33). Further evidence for Boulder Clay filling this glacial channel is recorded south of The Howe [812 318], where Boulder Clay occurs as a tongue extending from the plateau level at about 70 m down to below 40 m OD in the valley.

In the sand and gravel quarry at Alphamstone, about 7 m of stiff pale brown structureless chalky Boulder Clay rest with a sharp junction on sand and gravel. This quarry was closed down in 1975 and has been recently regraded; there are now no sections visible.

The sand and gravel quarry (Ferriers Pit) south of Speck's Farm [892 345] exposed Boulder Clay sandwiched between Glacial Sand and Gravel above and Kesgrave Sands and Gravels beneath; its top and base are sharp and there is no evidence of reworking. It thickens westwards from less than 0.5 m to over 6 m at the western limit of the pit. The base is almost horizontal, and the deposit thickens at the expense of the Glacial Sand and Gravel which is not present at the western limit [8927 3426]. The Boulder Clay consists of orange-brown pebbly and sandy clay with specks of chalk but, where it thickens in the west, it gradually becomes grey at the base and the chalk content increases. Where the Glacial Sand and Gravel is absent in the west of the pit, the uppermost 3 m of Boulder Clay is weathered to a pale yellowish brown hue, and much of the contained chalk is finely comminuted.

The base of the Boulder Clay falls from 60 m OD to below 40 m OD in the tributary valley running north from the River Colne near Poole's Farm [942 274], and then rises to 53 m OD (see Figure 12). There is a subsidiary fall into the major valleys, notably at Mount Bures [906 327] where the base falls steeply from 60 m OD beneath the plateau to below 40 m OD. The maximum recorded thickness of the Boulder Clay is in excess of 18 m in two

boreholes [9391 2927; 9541 2954] (Ambrose, 1974). The Boulder Clay is typically chalky over the whole area except in the east, where it thins towards its contact with the Glacial Sand and Gravel and becomes decalcified in places, for example in two boreholes [9658 3060; 9754 3467] (Hopson, 1981). A further area of decalcified Boulder Clay occurs [951 281] around West Bergholt Hall where an orange-brown clayey sand and gravel was augered; a nearby borehole proved 3.6 m of sandy clay with gravel overlying Kesgrave Sands and Gravels.

The River Colne to the Roman River, east of Great Tey

This area lies at the eastern edge of the Boulder Clay outcrop. On the crest of the interfluve and in the northern and eastern parts of the outcrop, the Boulder Clay is chalky at the surface and there are numerous small disused marl pits. From its eastern limit it thickens to around 10 m, although one borehole [9051 2650] proved chalky Boulder Clay from 2.7 m to 18.3 m where the boring was terminated (Ambrose, 1974, p.24). South of the road running from Checkley's Farm [899 262] to Aldham [917 257], the Boulder Clay is, in general, decalcified at the surface, and an ochreous brown sandy clay was augered beneath a flinty clay soil. Tongues of decalcified Boulder Clay [e.g. 915 249] also extend into the Roman River valley. Decalcified Boulder Clay was proved in two boreholes [9067 2562; 9171 2541] which penetrated 2.7 m and 4 m respectively of brown sandy clay (Ambrose, 1974).

A small outlier of Boulder Clay [915 245] lies in the floor of the Roman River valley south-west of Aldham Hall. This is the surface expression of a thick plug of Boulder Clay which occupies a deep channel beneath lacustrine deposits which are described in more detail on p. 42. Marks Tey No.2 Borehole sited on the outcrop, proved 19.7 m of grey chalky Boulder Clay, passing down into laminated grey clay at 9.1 m above OD. Chalky Boulder Clay underlies much of the lacustrine deposits of the Roman River valley, as is borne out by BGS trial pits and boreholes put down during the survey. A trial pit [9251 2416], west of Copford Place, was sited on lacustrine deposits which are interbedded with the Boulder Clay:

	Thickness m
Clay, brown, chalky from 0.9 m; decalcified, orange and pale grey with red mottles from 1.25 to 1.30 m; some subangular flints	1.55
Silt, clayey, interbedded with silty medium-grained sand, laminated, orange and pale grey mottled; chalk pellets throughout; minor micro-faulting	0.95
Clay, sandy, chalky with subangular flints; some lenses of chalky medium-grained sand	0.65
Silt, clayey and silty clay, finely laminated pale brownish grey, with beds of fine- to medium-grained orange-brown chalky sand	0.55
Clay, sandy, with chalk pellets and a few flint pebbles; very sandy in top part	0.65

Chalky Boulder Clay has been proved at quite shallow depth beneath lacustrine deposits around the edge of the lake basin; for example in a BGS trial pit at Lampitts Farm [9319 2500], which exposed 2.8 m of lacustrine silts (see p. 47) overlying grey chalky Boulder Clay. Boulder Clay has also been proved beneath thick lacustrine sediments in Marks Tey No.1 and No.3 boreholes. Complete logs appear in Appendix 1.   RAE

North of the River Stour

The base of the Boulder Clay falls from 61 m OD around Sawyer's Farm [907 372] to about 53 m OD to the north-east of Nayland. Its junction with the Kesgrave Sands and Gravels crops out without interruption along the River Stour valley and its tributaries at

Leights, ranging from about 53 m OD, near Bures to 46 m OD at Harper's Hill, Nayland [9636 3479]. The greatest thickness of chalky Boulder Clay is to be found under the highest parts of the interfluves, where the maximum thickness proved is 10.5 m in one borehole [9022 3701]. In general the Boulder Clay thins towards the Stour valley and also towards the eastern margin of the district, where between 1.3 and 4.3 m are recorded (Hopson, 1981).

The Boulder Clay is generally decalcified in the top one or two metres, except near its eastern margin where there is an orange-brown and grey sandy clay up to 3 m thick with subangular to rounded flint pebbles, for example in a borehole at Radley's Farm [9547 3525]. A small section at the top of Harper's Hill, Nayland, exposed a similar lithology resting on Glacial Sand and Gravel and the Boulder Clay passes laterally into the Glacial Sand and Gravel in this area (see p. 33).   RAE

GLACIAL SILT AND GLACIAL LAKE DEPOSITS

The deposits classified as Glacial Silt typically comprise well-laminated silts with subordinate interbedded clays and fine-grained sands. In the unweathered state these sediments are commonly pale grey, but at the surface they are generally cream or buff. The silts are typically calcareous and locally contain nodules of 'race' in the weathered zone.

In Essex as a whole, these sediments, which reflect sub-aqueous slackwater conditions, occur as lenses within both the Boulder Clay and the Glacial Sand and Gravel, and also locally, as comparatively thick sequences in the buried channel systems (see p. 16). In the last instance, some of the sub-glacial scours became lakes, when the ice melted, and some of the fine-grained glacial deposits can be confidently classified as Glacial Lake Deposits in their lower parts, passing upwards into interglacial Lacustrine Deposits, as for example at Marks Tey (see p. 40). In the Colne buried valley a mean thickness of 4.5 m was recorded for the Glacial Silt (Marks and Murray, 1981, p.7).   RDL

Details

At 1.6 km west of Gosfield Hall, Whitaker and others (1878, p.48) noted that a brickyard [759 299] exposed bedded clay, the upper part weathered brown and sandy, the lower bluish grey. These lithologies apparently occur close to the base of the Boulder Clay. Soft laminated silty clays at this level were recorded in a borehole [7933 2852] south-east of Gosfield.

The railway-cutting [7730 3621] near Newman's Farm exposed the following section:

	Thickness m
Gravelly Head (in two divisions, see p. 52) and fill	0.5 to 1.5
Buff to creamy white interlaminated silts, clayey silts and fine sand	up to 1.5
Chalky sand and gravel, cross-bedded, passing laterally into chalky coarse sand	c.0.5

RDL

Glacial Silt has been mapped principally on the west side of the River Brain between Braintree and Rayne. The deposits consist predominantly of cream and grey clayey silts, and locally [7475 2202] contain beds or lenses of chalky Boulder Clay. The occurrences south of the railway line appear to have a base at about 56 m OD, and extend up to about 64 m OD. London Clay is exposed on the lower flanks of the valley. North of the former railway line much of the outcrop is built over, and stratigraphic relationships are not so clear. The deposits appear to extend down to Alluvium level, and the upper limit of the clays, marked by a fairly sharp feature, is at about 55 m OD. In this area the deposits consist of yellow and grey chalky silts.

At a site 'on the southern side of the high road about a mile westwards of Braintree Church' close to Clap Bridge [741 229], Whitaker and others (1878, p.66) recorded laminated clay and loam with broken shells, and noted that the bed was similar to that in the valley of the Blackwater. The beds at this latter locality formed in the Hoxnian Interglacial (Bristow, 1985). The presence of shells in the Braintree deposits suggests that they, too, may be interglacial in age, but until their relationships to the Glacial Silt are known no certain conclusion can be drawn.

South of Perry Childs Farm, a borehole [7289 2444] on the east side of the Brain valley, proved 2.4 m of 'grey silt' beneath 10.7 m of soil and brown silty clay with flints [?Boulder Clay]; the silt rests on London Clay (Clarke and Ambrose, 1975).   CRB,RDL

CHAPTER 6

Quaternary Drift: Hoxnian to Flandrian

When the Anglian ice was at its maximum a lake formed adjacent to the London Clay ridge at Messing and terrace gravels (Terraces 4 and 5) were deposited as deltaic fans in the lake, recording successive falls in the lake level.

A similar lake formed at Marks Tey when the ice melted and was so deep that it continued to fill during the succeeding Hoxnian interglacial stage, providing one of the thickest British sequences of this age, and one that established the local stratigraphical position of the lake silts.

The subsequent Quaternary history is represented by only minor deposits, whose precise age is far from certain. The Terrace 1 and 2 gravels of the main river systems accumulated much later than the Hoxnian although they have almost certainly reworked, at least in part, the earlier sands and gravels. The terraces were aggraded during a period of high sediment discharge in a relatively cold period when a limited vegetation cover allowed a considerable amount of erosion. A further factor influencing the formation of terraces, particularly in the lower reaches of the Stour and Colne river basins, was the fluctuation of sea-level, which, was at about 8 m above OD during Ipswichian time. It seems on balance that the Terrace 1 and 2 gravel was finally aggraded during the Devensian (post 40 000 to 50 000 years BP) and that the present surfaces are erosional benches cut across the deposits. Downcutting and reworking of these same gravels has continued until the present day, with channels cut at least 5 m into them and then filled with silts and clays containing organic debris. During the last 10 000 years or so, river discharge has, however, been insufficient to rework any substantial quantities of gravel, and wide alluvial flood plains have developed in the main river valleys.   RAE

LACUSTRINE DEPOSITS

Whitaker and others (1878) described interglacial deposits in the Blackwater valley only 1 km south of this district. They were mapped in the 1960's, and described by Bristow (1985) who found they were associated with the Third Terrace of the Blackwater. Turner (*in* Rose and Turner, 1973) studied borehole material and temporary sections in the deposits and concluded that they contained pollen indicating a Hoxnian age.

The deposits have been penetrated in two boreholes in the present district although their outcrop is not separately delineated, since they have been included within the outcrop of Terrace 3 deposits. One borehole [8556 1829] proved soil, 1.5 m; yellow and grey plastic clay, 2.5 m; grey, blue and light green silts with wood fragments in the upper part, 10.9 m; overlying more than 9.5 m of sand and gravel [Terrace 3] (Haggard, 1972, p.46). Some 0.6 m of calcareous creamy silts were augered [8666 1823] 130 m south-east of this borehole. The extension of these Lacustrine Deposits north-eastwards is indicated by the presence of marl on the

surface of an old trench [8588 1852] and by the presence of the lime-loving *Clematis vitalba* in a hedgerow [8615 1864] on the gravelly soil of allotment gardens nearby. Another borehole [8710 1926] proved 1.7 m of top soil and gravel, overlying 2.3 m of laminated clay (presumed Lacustrine Deposits) which in turn overlay 6.4 m of sandy gravel [Terrace 3] (Haggard, 1972, p.5). This borehole was sunk alongside an old pit, augering in the banks of which revealed up to 1 m of greyish brown silty clay. *Clematis vitalba* was noted in a hedgerow [8713 1940] just north of Threshelfords Farm, and may indicate a further extension of the laminated clay. An occurrence of 2.1 m of grey silty clay with race was also noted in a ditch [8635 1985] north of Kelvedon, where a thin spread of sandy and gravelly Head overlies Terrace 3 deposits.   CRB

Much more significant outcrops of Lacustrine Deposits have long been known in the vicinity of Marks Tey [910 243] and Copford Place [933 242] where they have been excavated for brickmaking. The deposits at Copford were described in detail by Brown (1843; 1852) who noted their flora and fauna. Subsequently, in the period 1873 to 1875, the area was mapped by W. H. Dalton of the Geological Suvey who assigned all the lacustrine deposits to 'Post Glacial strata' in his memoir (Dalton, 1880). Since about 1863, bricks have been made by Messrs W. H. Collier at the Marks Tey Brick and Tile Works [912 243] some 400 m north of Marks Tey Church. Turner (1966; 1970) undertook an exhaustive stratigraphical and palynological study and the following account is, with his permission, based largely on his doctoral thesis (Turner, 1966).

The roughly rectangular outcrop of the deposits extends eastwards on either side of the Roman River for about 3 km south from Church House Farm [905 245] to Stanway village [935 242]. The deposits probably also occur farther west beneath Head on the south side of the Roman River, and locally beneath reworked Glacial Sand and Gravel on the north bank. A further small outcrop lies west of Teybrook Farm [898 250]. The main outcrop is broken by patches of Boulder Clay [916 245] and Glacial Sand and Gravel [923 247] whose relationship to the Lacustrine Deposits is discussed below. The recent alluvium of the Roman River crosses the outcrop diagonally and then follows its southern boundary to Stanway.

Most of the outcrop lies within a shallow depression, which is bounded by a weak break of slope where it abuts against Boulder Clay on the western and southern margins and London Clay and Glacial Sand and Gravel to the north and east. The general surface level of the depression falls eastwards from about 32 m to 25 m OD and its topography is subdued, except where it is broken by a rounded hill of Glacial Sand and Gravel near Hole Farm.

Glacial deposits, probably mostly Boulder Clay but with interbedded Glacial Silt and Glacial Sand and Gravel, everywhere underlie the Lacustrine Deposits, and together

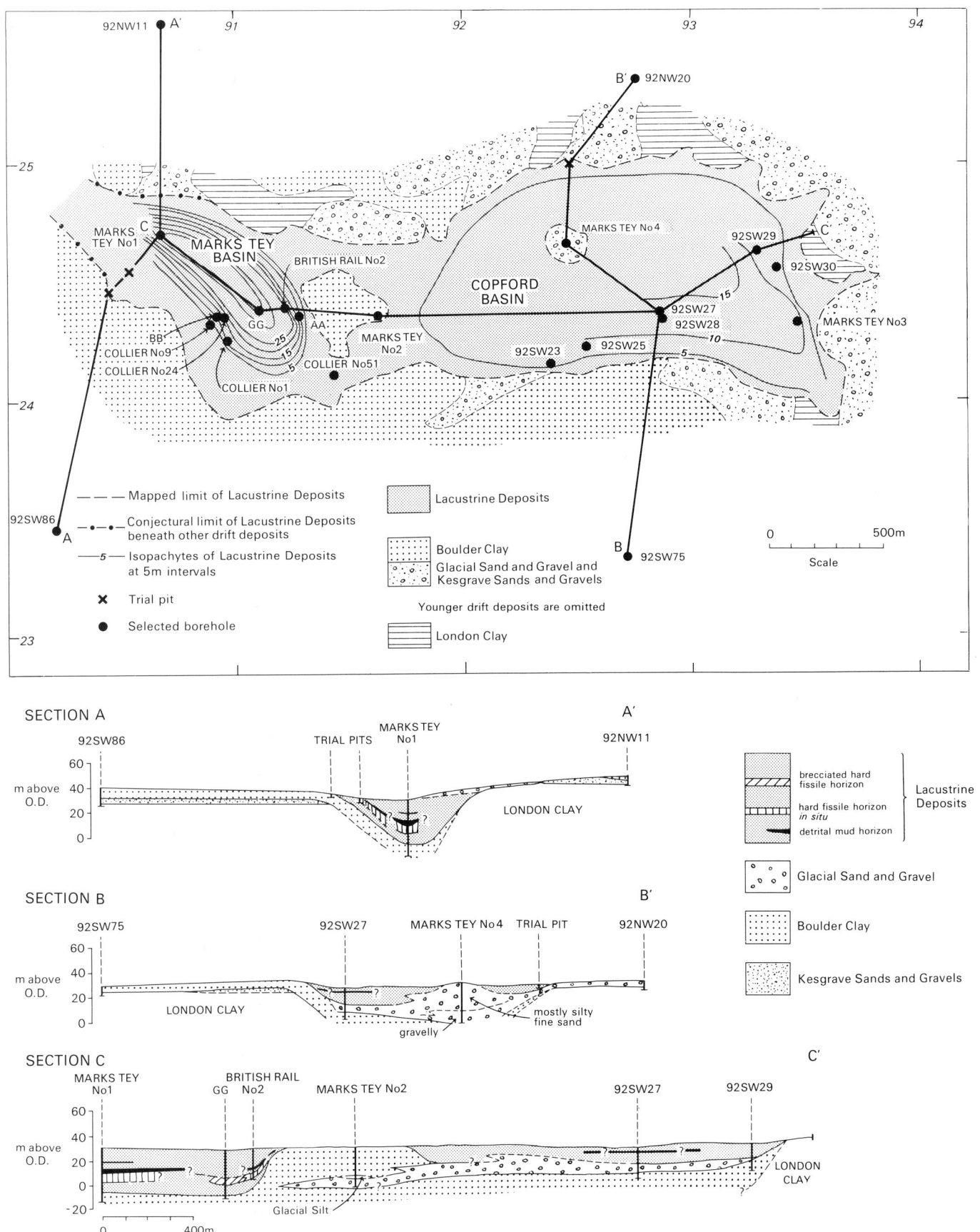

Figure 14 The drift deposits at Marks Tey

these sediments fill a deep depression which was probably excavated below standing ice (Figure 14). Turner (1970) demonstrated from borehole records that the Boulder Clay of the adjacent plateau passes beneath the Lacustrine Deposits. Nowhere in the depression has any borehole passed through Lacustrine Deposits and glacial deposits into bedrock; the deepest boring, Marks Tey No.1 [9066 2473], terminated at 11.6 m below OD within glacial deposits. The bedrock is likely to be London Clay down to about 15 to 20 m below OD; the top of the Chalk probably lies around 35 to 40 m below OD.

The Lacustrine Deposits occupy two basins. These are effectively separated by a mass of glacial deposits which come to the surface to form an outcrop of boulder clay [916 245]. Marks Tey No.2 Borehole (see Appendix 1) was sited on these glacial deposits which are overlapped by thin Lacustrine Deposits to the north and south (see Figures 14 and 15). The western basin is here informally termed the Marks Tey Basin and the eastern, the Copford Basin. Available borehole evidence suggests that the former basin is the deeper (Figure 14), but their morphology is imprecisely known. Nevertheless there is no direct evidence of any outfall from the Marks Tey Basin into the Copford Basin or from the Copford Basin towards the south-east. The postulated shape of the Marks Tey Basin, as shown in Figure 14, is determined largely by the two deepest borings, namely

Marks Tey No.1 and (Turner's) Borehole 'GG', which respectively proved Lacustrine Deposits down to 5.3 and 2.9 m below OD, giving complete thicknesses of 34.5 and 35.0 m. By comparison, the Copford Basin is thought to be less deep, as a maximum thickness of only 14.5 m of Lacustrine Deposits, whose base is at 11.5 m above OD, has been proved.

The feather edge of the Lacustrine Deposits lies at 30 m above OD only some 250 m away from Marks Tey No.1 Borehole which implies that the average gradient of the base of these deposits is here 7° to 8°. An even steeper slope of 14° can be deduced between boreholes AA and BR2 (see Figure 15). The relatively steep sides of the basin may have caused mass movement within the accumulating sediments (see below). On the lip between the two basins the lowest proved level of the Lacustrine Deposits lies at about 28.4 m above OD, and they occur up to 33 m above OD.

At the surface the deposits are characteristically soft, laminated, pale buff and grey mottled, clayey silts and clayey fine-grained sands, which contain small thin-shelled gastropods and calcareous 'race' nodules up to 1 cm in diameter; thin beds of chalky silt and chalky sand also occur. In places a veneer of clayey gravel, which is locally as much as 2 m thick, obscures the Lacustrine Deposits; these are Head deposits which are not illustrated on the geological map because their extent cannot be mapped accurately.

The only existing large exposure of these beds is in the pit

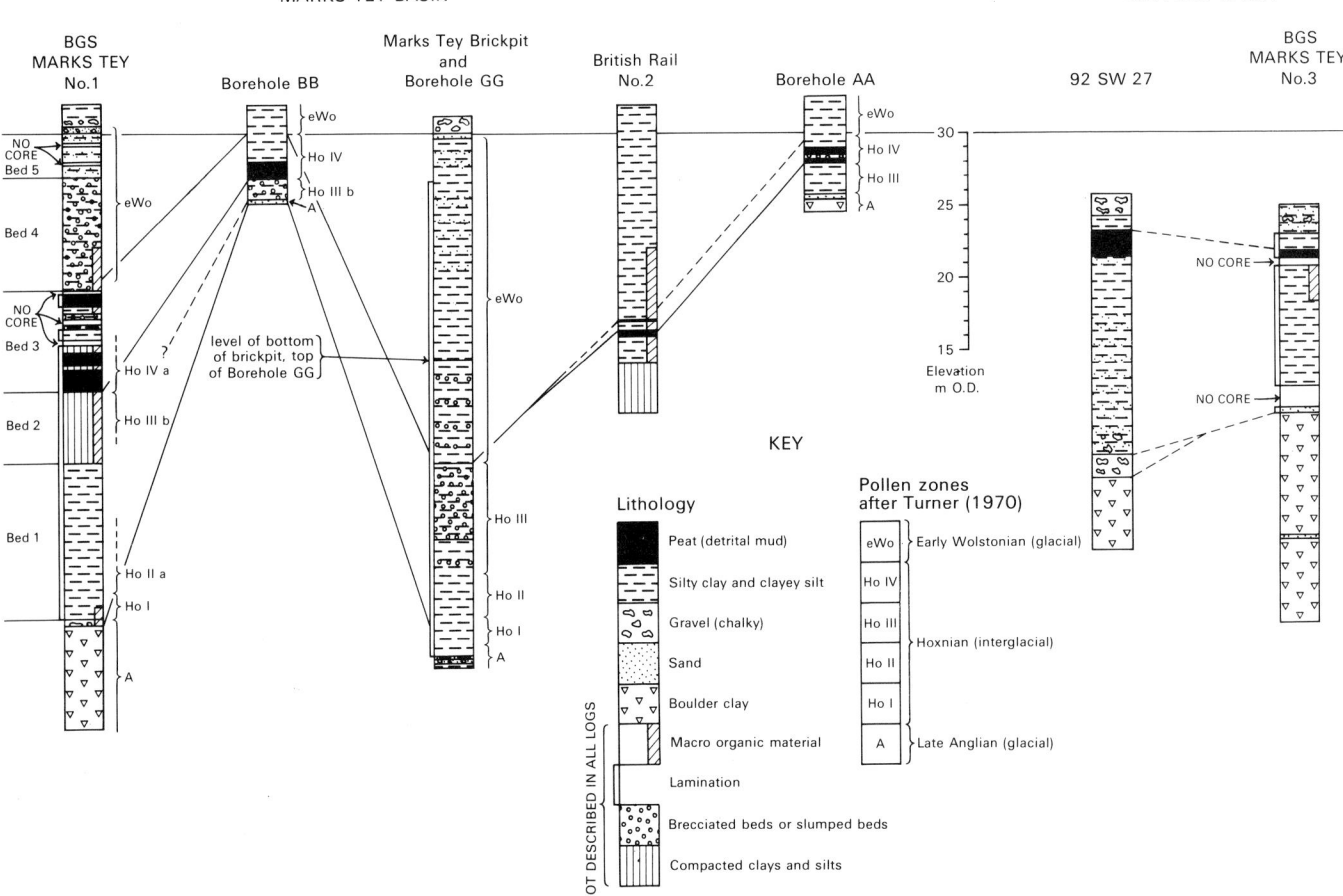

Figure 15 Comparative borehole sections in Lacustrine Deposits

at Marks Tey brickworks [9087 2447] (see below). Borehole GG was put down in the bottom of this pit, and the stratigraphy was described in great detail by Turner (1966; 1970); a simplified graphical log is given in Figure 15, and can be compared with that of Marks Tey No.1 Borehole, drilled about 550 m to the north-west.

The pollen zonation provided by Turner for these and other boreholes is also indicated in Figure 16. The vegetational history has been traced from the late Anglian (glacial) stage (Zone A) through the Hoxnian (interglacial) (Zones HoI to IV) to the early Wolstonian (glacial) stage (Zone eWo) (see below). The deposits exposed in the workings all belong to the early Wolstonian.

Marks Tey No. 1 Borehole has provided a reasonably complete section through the Lacustrine Deposits close to the deepest part of the Marks Tey Basin. Accordingly it has been used as a convenient reference section to which other sections in the Lacustrine Deposits can be correlated, and the strata have been divided informally into five 'beds', numbered from below upwards. The full log of this borehole is given in Appendix 1 and a simplified log is given below:

	Thickness m	Depth m
Head		
Silt, orange brown and grey mottled, with sand and gravel at the base	2.00	2.00
Lacustrine Deposits		
Bed 5: Silt, clayey, and clay, soft, dark grey to greenish grey, with race and chalk pellets	3.14	5.14
Bed 4: Silt, clayey and clay, silty, soft to firm, grey, greenish grey and olive grey, commonly finely brecciated; scattered shells and shell debris; organic remains in the basal part	7.76	12.90
Bed 3b: Silts, organic, detrital, brown, fissile; shown as 'peat' in Figure 15	1.10	14.00
Bed 3a: Silt, clayey, microbrecciated, interbedded with organic silt and finely laminated silt	6.00	20.00
Bed 2: Silt, brown, fissile, hard, leathery, overconsolidated	5.00	25.00
Bed 1: Silt, clayey, and clay, silty, dark grey to greenish grey; finely laminated	11.40	36.40

Although rhythmically bedded lithologies occur commonly within this and other sequences, brecciated horizons are also present (Figures 14 and 15), and the effects of sediment recycling are difficult to assess.

The Lacustrine Deposits are thinner towards the margins of the former lake where they include firm detrital muds and brecciated beds. Only the higher pollen zones of the Hoxnian and early Wolstonian (Ho IIIa to eWo) have been recognised.

Bed 1 in Marks Tey No.1 Borehole and the beds from 12.54 to 20.87 m in Borehole GG are both finely colour-laminated silt, the lamination reflecting rhythmic variations in organic content. In this interval Turner (1970, pp.391–395) has recognised alternating calcareous-rich and carbonaceous-rich laminae, and suggested that each pair may represent an annual accumulation of sediment, the calcareous laminae representing Spring-time diatom flushes in water 25 to 30 m deep, and the carbonaceous laminae being summer, and particularly autumn, accumulations. From 20.12 to 20.22 m depth in Borehole GG, Turner (1970, p.434) recorded a restricted non-marine molluscan fauna of

Bithynia tentaculata, *Valvata piscinalis* and *Pisidium sp.*, indicating conditions of gently flowing water.

Bed 2 of Marks Tey No.1 Borehole, where it may be regarded as overconsolidated, has been correlated on pollen evidence with the interval from 7.17 to 12.54 m in Borehole GG, which was described by Turner (1970, p.385) as 'an irregular jumble of angular brecciated blocks of highly fissile laminated clay mud'. This breccia probably represents a slumped mass of unstable sediments that were originally lithologically similar to those of Bed 2 in Marks Tey No.1 Borehole.

In a trial pit [9052 2457] at the western edge of the outcrop a brecciated bed of dark brown micaceous silt, thought to be equivalent to Bed 2, lies *in situ* beneath 0.4 m of organic detrital silt with large wood fragments and twigs, thought to be equivalent to Bed 3. This peaty bed, which should more correctly be termed detrital organic mud as there are no basal roots recorded, forms a convenient marker which occurs widely in both the Marks Tey and Copford basins (see Figure 14). The absence in Borehole GG of any of the detrital organic muds that lie within Bed 3b in Marks Tey No.1 Borehole (equivalent to Zone Ho IV) suggests a possible hiatus in deposition in the former borehole but it is equally possible that continued recycling of sediment modified the resultant pollen spectrum.

Recycling of sediment certainly took place within Bed 4 of Marks Tey No.1 Borehole, though not in equivalent strata in the brickpit; the strata in both sections yield similar pollen spectra of early Wolstonian aspect.

Marks Tey No.3 Borehole provides the best reference section of the Copford Basin, although it lies near its eastern margin (Figure 15):

	Thickness m	Depth m
Head:		
Sand, clayey and silty, with flint grave	2.00	2.00
Lacustrine Deposits:		
Silt, calcareous, banded pale and dark grey, laminated	1.23	3.23
Mudstone, organic, black, fissile, highly fissured	0.37	3.60
No core	0.60	4.20
Silt, clayey, grey and greenish grey thinly laminated, with shells and some chalk pellets	8.30	12.50
No core	c.1.50	c.14.00
Glacial Sand and Gravel:		
Sand, silty, fine- to medium-grained, grey, with chalk fragments	0.30	14.30

This succession appears broadly comparable with that in the Marks Tey Basin. No equivalent of Bed 2 was, however, recorded from borings in the Copford Basin, but a lithology similar to that of the brecciated beds of Borehole GG was recorded 2.7 m below ground level in a trial pit dug during the present survey west of Copford Place [9251 2417], where rafts of laminated dark grey-brown (5YR 3/1) and olive (5Y 3/1), micaceous, organic, fissile clay were inclined at angles of up to 45° and set in a matrix of soft, medium grey (N4) and brown mottled clay.

The equivalent of the organic mudstone between 3.23 to 3.60 m in Marks Tey No.3 Borehole is recorded from many of the trial borings in the Copford Basin. If this lithology is equivalent to Bed 3 of the Marks Tey Basin, the upper part of the sequence must be attenuated, only about 5 m being present at most. This part of the succession is locally exposed in the banks of the Roman River [934 242] where it consists of soft, pale grey and brown mottled, clayey silt with some small chalk pellets and thin interbeds of coarse chalky sand and calcareous silt.

On the southern margin of the Copford basin, old brickworkings once exposed the top part of the lacustrine sequence (Brown, 1852; Dalton, 1880). The exposures showed calcareous shelly silts overlying a compact detrital mud with *Valvata piscinalis* which in turn overlay laminated clay with freshwater shells, including *Valvata piscinalis* and *Bithynia sp.* opercula (see p. 47 for details). Sixty-nine species of freshwater and terrestrial molluscs, along with teeth and bones of elephant, stag, bear and beaver, were recorded from the topmost strata by Brown (1843; 1852). Turner has recently reexamined the fauna and processed samples originally collected by Brown. A pollen spectrum obtained from organic silt retained within the shells has led him to conclude that these deposits, from the uppermost 2 m or so of strata in the Copford Basin, are of Flandrian age. These beds are commonly recorded in borings as 'white chalk' or 'marl' and they probably represent a band of highly calcareous clay which has been oxidised but not decalcified.

Marks Tey lies at the eastern limit of the Anglian ice-sheet where a steep-sided tunnel-valley is thought to have been excavated perpendicular to the ice-front, by subglacial water at high hydrostatic pressure. As the ice-sheet waned towards the close of Anglian times, the deep trough became partly filled with meltout sediments and laminated Glacial Silt deposited subglacially or in small ice-dammed lakes. When the trough was vacated by the ice a more extensive lake formed, which was possibly initially dammed by masses of stagnant ice, and laminated argillaceous sediments derived from the adjacent glacial deposits were rapidly deposited in the lake. One large block of 'dead' ice occupied the central part of the trough whilst a further dam may have blocked the exit in the east. A large Boulder Clay plug remains as evidence of the central dam, but there is no evidence of a major outlet or a dam of glacial sediments to the eastern end of the trough. The lake persisted throughout the Hoxnian stage, and was filled with sediment during this interglacial and the subsequent Wolstonian stage.

Turner (1970), concluded that, except for a short period in the middle part of its history (HoIII and IV), the Marks Tey Basin drained into the Copford Basin, which in turn had an outlet along the present Roman River valley. The latter, although it is incised about 20 m through Kesgrave Sands and Gravels into London Clay, contains no evidence of the age of incision nor does it contain any glacial outwash deposits. The lowest part of the sequence in the central parts of the Marks Tey and Copford basins (Bed 1) accumulated in quiet water that permitted uninterrupted sedimentation. No marginal lacustrine sediments are preserved, but it is reasonable to assume that there were peaty beds and small delta fans at the mouths of stream channels.

During the middle part of the Hoxnian, the water-level in the lake fell by about 5 to 10 m to about 20 m above present OD, resulting in the emergence of the central area of Boulder Clay that divided the lake into two. The reason for such a drop is not known. It is probable that two separate lakes lay within the two closed basins because their outfalls could have been no lower than 20 to 25 m above present OD. Because of the drop in water-level and consequent

emergence of earlier lake sediments around much of the lake, the uppermost parts of the sediment pile were partially dewatered and consolidated to a hard leathery consistency (Bed 2), and were locally desiccated and cracked to produce the autochthonous breccias noted in boreholes and trial pits (pp. 46 and 62). Borehole GG records a subsequent period of slumping when the 'Bed 2' brecciated beds became unstable and slumped into deeper water: Turner (1970) was able to recognise the basal shear plane of the slumped mass at 12.51 to 12.54 m depth in Borehole GG (see p.64). Plant colonisation of the exposed basin-floor was rapid during the period of low water-level, which coincided with the climatic optimum of the Hoxnian. The lake water-level then rose again, probably in respose to heavier rainfall. Dead vegetation was redeposited as prominent detrital mud bands (Bed 3) which are preserved in both the basins. Overlying interbedded deeper-water laminated silts, brecciated beds and detrital muds suggest small fluctuations in a gradually rising lake-level which ultimately again reached to about 30 m above present OD. During this period there was a further accumulation of lacustrine laminated clayey silts in the centre of the Marks Tey Basin, though these silts thin rapidly towards the margins (Beds 4 and 5).

The fine brecciation of Bed 4 in the Marks Tey No.1 Borehole is difficult to interpret. There is little coarse clastic sediment and, therefore, reworking by vigorous streams flowing in from the north seems unlikely. The same levels in Borehole GG are not affected, but settlement of the sediments has produced minor normal faulting throwing down towards the centre of the basin.

The topmost part of the lacustrine sequence consists of calcareous silts, containing freshwater shells (*Valvata piscinalis*, *Bithynia tentaculata*) characteristic of slow moving water. These chalky sediments, which become coarser upwards, were probably brought into the basins, following increased erosion of the surrounding glacial deposits, in a climate which was deteriorating at the onset of the Wolstonian cold period. The lower annual mean temperature reduced vegetation cover, and freeze-thaw action probably mobilised solifluction flows which were carried into the lakes. The final lake filling is a coarse chalk and flint gravel solifluction deposit with ice-wedge casts, marking the onset of permafrost conditions during the Wolstonian.

Throughout the progressive filling of the two basins, local streams brought in fine-grained sand with minor amounts of coarse sand and gravel from the surrounding slopes. The relationship of these deposits to the lacustrine deposits of the Copford Basin is not clear, but in Marks Tey No.4 Borehole they are here interpreted as 'deltaic' (see Figure 14).

The oldest pollen-bearing lacustrine sediments were laid down in late Anglian times when the temperature increased rapidly and birch (*Betula*) quickly colonised the ground vacated by the ice-sheet (Turner, 1970). The area was predominantly open grassland with sedges and willows in marsh hollows and along spring-lines. By the beginning of the Hoxnian (Ho I), birch woods were widespread, and pine (*Pinus*) immigrated into the area. A mixed oak forest with alder (*Alnus*), hazel (*Corylus*), ivy (*Hedera*) and holly (*Ilex*) subsequently became established in a temperate climate (Ho II). During Ho III there was a progression to a mature forest with the incoming of hornbeam (*Carpinus*) and the silver fir (*Abies*), and a decline of oak (*Quercus*) and elm (*Ulmus*), indicating degeneration of the forest soils, although the

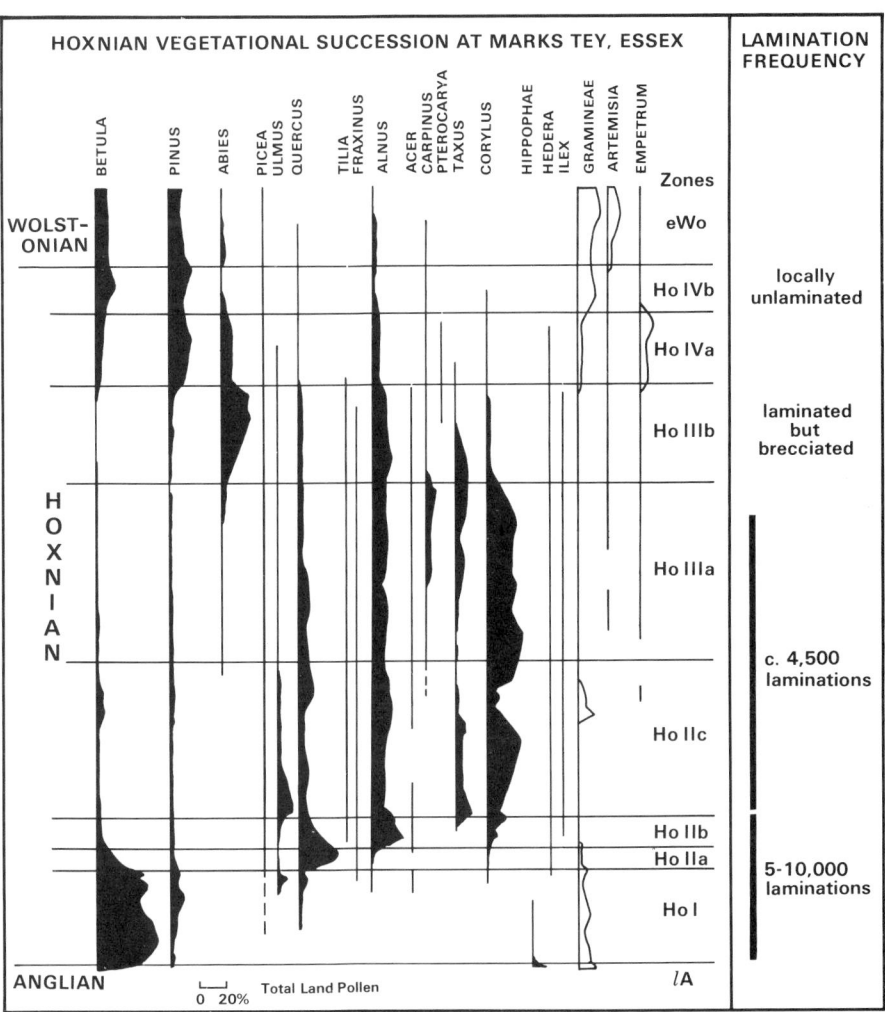

Figure 16 Hoxnian vegetational succession at Marks Tey (composite pollen diagram; Turner, 1973)

presence of box (*Buxus sempetuirens*) and vine (*Vitis*) shows that high summer temperatures were maintained.

When the climate deteriorated after the thermal maximum of the interglacial, the mixed woodlands gradually died out, to be replaced (Ho IV) by more open country with *Empetrum* heathlands and birch and pine woods. The onset of much colder weather precursing a further glacial stage is indicated by the spread of *Artemisia*, and only low-growing open vegetation existed in the permafrost conditions that followed in much of the Wolstonian, although birch and pine are represented in the early sediments (Wo Ia) at Marks Tey.

There is no reliable radiometric date for these Hoxnian deposits, although fossil bones from the Hoxnian at Clacton (Szabo and Collins, 1975) have been dated at about 220 000 to 280 000 years BP. Assuming that there are two major cold periods after the Hoxnian and that sediments from each major cold and warm period are represented in Britain, several authors have equated the Hoxnian with Stage 7 of the oxygen isotope curve, which is put at about 240 000 years BP by Kukla (1977) and 251 000 years BP, by Shackleton and Opdyke (1973).

By counting annual laminae in the Ho I and Ho II zones and by estimating the rate of sedimentation in the other Hoxnian zones, Turner has postulated the duration of the entire Hoxnian at Marks Tey to be in the order of 20 000 to 25 000 years.

Several radiocarbon dates have been obtained from samples of Lacustrine Deposits in the general Marks Tey area. A bed of detrital organic material, which is exposed in a ditch beside the A12 road [9257 2425] in the Copford Basin, has been dated at 32 500 years BP (sample St.3864) and 28 170 years BP (sample St.3846). A further date obtained from an organic horizon at the margin of the Marks Tey Basin [9052 2457] gave a date of 38 724 years BP (sample SRR1063). It has been generally agreed (Bristow and Cox, 1973) that such dates, which are referable to the middle part of the Devensian stage, but are recorded from samples taken within 3 m of the surface, are likely to be erroneous because of contamination by percolating humic acids in groundwater. By comparison, dates of greater than 40 000 years BP were obtained from two, presumably uncontaminated samples from 13.8 to 14.0 m (sample SRR 1066) and 18.0 to 18.5 m depth (sample SRR 1065) in Marks Tey No.1 Borehole. The marginal Lacustrine Deposits from which the finite radiocarbon dates have been obtained are thought to be Hoxnian (probably zone Ho IV); if so the dates are either false or the entire lacustrine sequence is much younger than the date generally proposed for the Hoxnian. RAE

Details

South of the Roman River and to the west of the railway the upper part of the Lacustrine Deposits are documented in numerous shallow borings put down by W. H. Collier, and the same parts of the sequence are also exposed in the working brickworks. A trial pit put down near the western extremity of the outcrop [9052 2457] proved, beneath 1.0 m of Head, the following:

	Thickness m
Lacustrine Deposits:	
Mottled orange and grey (10YR 6/6 and 5Y 6/1), medium- to coarse-grained, sandy clay with angular to subangular flints up to 2 to 3 cm diameter and small rounded quartzite pebbles up to 5 mm across; race nodules throughout; a lens of grey to buff coarse-grained sand with a clay matrix and some flint gravel up to 3 cm diameter, from 0.40 to 0.55 m; gradational base	1.10
Soft sticky dark grey brown clayey silt to silty clay, with thin silt streaks and laminae and small mica flakes; lens containing 1 mm diameter chalk pellets at top of bed; gradational base, with increasing organic material	0.15
Laminated dark purplish brown (5YR 2/2) organic clayey silt with some greenish brown mottling; pale buff discontinuous silt laminae and some finely comminuted shell fragments	0.55
Detrital organic material with large (10 mm diameter) wood fragments; chalk pellets in thin silty lenses up to 2 to 3 cm thick, and comminuted shells near top of bed	0.40
Hard micaceous dark brown clayey silt with some subangular flints; beds are broken or brecciated	seen to 0.30

The topmost beds of the sequence are exposed in old, partly overgrown, faces in the brick-pit [9087 2447], about 500 m north-west of the works:

	Thickness m
Head:	
Pockets of dark brown clayey sand and angular flint gravel with pebbles up to 10 cm across; cryoturbated base	0.20 to 1.50
Lacustrine Deposits:	
Soft medium to pale grey silty clay and clayey silt with some flint pebbles up to 3 cm in diameter; highly calcareous in places and faintly laminated; weathers to pale brown in patches	2.00 to 3.30
Massive medium to pale grey (N5 to N7) clayey silt with crushed thin-shelled bivalves	0.23
Pale grey-brown medium-grained sand with shell debris; uneven base	0.09
Medium grey clayey silt with harder brown partly iron-cemented bands; laminated near the top but massive below with fining-upwards sandy beds 3 cm thick; flint and chalk pebbles up to 1 cm diameter in basal 26 cm	0.86
Grey brown clayey silt with organic horizons and platy calcareous silt at base	0.15
Chalk grit, slightly clayey with interbeds of platy chalky siltstone pellets c.5 mm across with rounded edges; some organic material on bedding surfaces; beds generally less than 1 cm thick	0.16
Pale grey and brown mottled fining-upwards beds of chalky silt to coarse sand with platy silt interbeds	0.15

	Thickness m
Massive grey and brown mottled sandy clayey silt with some cyclic fining-upwards cycles generally less than 3 cm thick (poorly exposed)	seen to 3.03

The working brick-pit exposed about 6 m of finely laminated silty clays with alternating bands of medium grey (N4) and dark grey (N2 and N3) organic horizons. Some laminae contain 1 mm diameter chalk pellets and there are thin discontinuous layers of fine- to medium-grained sand near the top of the face. Small moss fragments and some thin-shelled molluscs are also seen. Normal faults, with throws of up to 10 cm, break the essentially continuous colour banding around the working face which is about 50 m in length.

North of the Roman River, the mapped northern boundary of the Lacustrine Deposits rises to its highest level, at around 35 m OD. The deposits in this vicinity, west of the minor road running from Aldham Hall [9185 2487], are generally chalky, and contain interbeds of poorly sorted gravel and some laminated chalky sands and gravel along with layers of chalky Boulder Clay. They are generally reworked from the adjacent glacial deposits and may, in part, be solifluction deposits contemporaneous with the upper part of the Lacustrine Deposits, a point illustrated by a trial pit [9251 2416] put down at an elevation of 34.1 m OD:

	Thickness m
Clay, brown, chalky from 0.9 m; decalcified orange and pale grey mottled horizon with rounded pebbles and subordinate angular flints together with belemnite and *Gryphaea* fragments from 1.25 to 1.30 m	1.55
Clayey silt with some sand grains, orange and grey striped, with some laminated greenish grey and orange silts and red-brown silty medium-grained sand interbeds; chalky throughout and dipping to the south (c.5°); local gravelly base with rounded flints	0.75
Clay, sandy and chalky, brown, with subangular flints and with chalky medium-grained sand lenses	0.65
Clayey silts, striped as above with pale brown and grey clays and orange-brown sand interbeds with chalk and race pellets; some microfaulting	0.65
Chalky sand, becoming clayey and passing into chalky sandy clay (Boulder Clay), with chalk pebbles and flints up to 4 cm diameter	seen to 0.55

East of Aldham Hall and north of the London to Colchester railway, the Lacustrine Deposits are poorly documented. Three pits have been dug into the marginal deposits during this survey. South of Chippetts [9242 2506], the following section was proved beneath 1.2 m of Head:

	Thickness m
Clay, stiff pale grey and brown-orange mottled; abundant carbonaceous remains including a dense concentration of roots immediately below the Head; becoming bedded with interbeds of fine sand and stiff clay; scattered flints up to 2 cm; sharp base	0.45
Silt, clayey and shelly with interlaminated silty clay; pale cream; more shelly bands, especially at 1.95 m; poorly defined graded base	0.65
Clay, mottled grey and brown, stiff	seen to 0.70

Lacustrine Deposits rest on Boulder Clay [9319 2500] at about 30 m OD, some 200 m north of Moat Farm:

	Thickness m
Top soil	
Clayey silt with scattered pebbles	0.40
Lacustrine Deposits	
Clay, sandy, pale brown with blue vertical veining and carbonaceous material throughout; a large clast of reworked London Clay at 1.80 m; in places a discontinuous pebbly horizon occurs above	1.20
Sand, fine-grained, with clayey bands	1.20
Boulder Clay	
Chalky sandy clay, grey	seen to 0.20

A tongue of Lacustrine Deposits immediately east of Moat Farm [933 248] rises to about 33 m OD where it rests on Glacial Sand and Gravel as proved in a pit [9347 2489]:

	Thickness m
Made ground	1.0
Lacustrine Deposits	
Firm mottled grey and dark orange slightly silty clay with abundant modern roots; small race nodules (up to 2 mm diameter) concentrated along grey veins	0.75
Clayey silt, grey and orange brown mottled, laminated; with fine-grained sand laminae many with chalk pellets and fine angular flint gravel	1.25
Glacial Sand and Gravel	
Coarse-grained sand, with fine to medium flint gravel and common chalk pebbles up to 3 cm diameter	seen to 0.20

South of the main railway line there have been numerous borings along the A12, although only a few penetrated the full thickness of Lacustrine Deposits. Selected borehole sites are shown in Figure 14. The logs of the holes are very generalised but indicate that, beneath a layer of weathered brown clay generally less than 2 m thick, a creamy chalky silt overlies a peaty bed at a depth of about 3 to 4 m. Similar strata crop out to the south of the A12 in a ditch [9255 2425], where they overlie 1 m of grey clayey silt with some broken thin-shelled gastropods.

The southern limit of the Lacustrine Deposits occurs about 2 m above a break of slope, where the deposits are presumed to overlie Boulder Clay. Two trial pits, dug within 20 m of each other and near the boundary, proved this contact to be irregular and complex. The more southerly [9251 2416] showed:

	Thickness m
Top soil	0.5
Head:	
Stiff dirty brown clayey medium to coarse sand, with common, mostly angular, flint gravel up to 5 cm diameter; gradational base	0.62 to 0.65
Stiff dark purplish brown organic clayey fine sand with scattered angular flints and rounded quartzite pebbles; gradational diffuse base	0.05 to 0.10
Lacustrine Deposits:	
Pale orange to brown laminated fine to medium sand and subordinate buff to pale grey clay beds up to 2 cm thick, containing thin sandy partings; the clay laminae have sharp tops and bases; stringers 1 to 2 cm thick containing rounded chalk pellets and a few belemnites; a	

	Thickness m
concentration of chalk pellets up to 3 cm diameter occurs at the top of the unit; microfaulting throughout	1.05
Glacial Sand and Gravel:	
Dirty brown chalky coarse sand and fine to coarse gravel; with chalk clasts up to 3 cm diameter and unworn flints up to 3 cm; the bed is cut out on a fault plane; sharp contact	0 to 0.10
Boulder Clay:	
Chalky stony clay, grey, plastic; affected by microfault; sharp base	0 to 0.60
Glacial Sand and Gravel:	
Laminated medium to coarse sand and yellow brown clay, with variable gravel, dirty brown and poorly sorted (compared to the unit above the Boulder Clay); the gravel consists mostly of chalk pebbles up to 3 cm across with subordinate flint and quartzite; becoming increasingly gravelly in lower part with no distinct lamination	up to 1.0
Glacial Silt:	
Laminated, pale yellow-brown, clayey silt	seen to 0.10

The second pit [9251 2417], dug on the Lacustrine Deposits, proved:

	Thickness m
Top soil and made ground — possibly old ditch filling	to 1.70
Lacustrine Deposits:	
Silt, calcareous, white, clayey in places and with shell fragments; lenses of well sorted, fine to medium sands, pale buff with shell fragments; becoming buff grey with depth, more clayey, and with fragmented shells; many modern roots	0.35
Interbedded sequence of clayey silts, slightly organic in parts with shell fragments, pale grey-brown; wispy bedding picked out by organic horizons, but no obvious laminations; beds up to 3 cm thick; sand bands tend to be fine at bottom and top but coarser in centre, and contain broken shells; the thickest bed is 13 cm	0.25
Clayey silt, pale brown, in places silty clay; shelly in parts and a trace of lamination; sharp base	0.15 to 0.35
Disturbed beds of laminated organic micaceous shaly clay, dark grey-brown to olive-black; inclined rafts up to 45° from horizontal within a soft slightly micaceous medium grey and brown mottled structureless clay	seen to 1.15 to 1.40

Some 100 m east of these trial pits is the site of the now disused Copford brickworks. The overgrown shallow workings contain cream calcareous silt and soft grey and pale brown mottled clays. The deposits were studied in the 1840s and 1850s by Brown (1852), who recorded a sequence similar to that described from the A12 borings and paid particular attention to its contained bones and shells. A generalised section taken from his work is as follows:

	Thickness m
Red brown clay with chalk, flints and other erratics derived from Boulder Clay	0.3 to 1.63
White shell marl with beds of ferruginous sand containing well preserved freshwater and land shells and bones of ox and elephant and stag antlers; dips to the north	0.3 to 1.83

	Thickness m
Very compact vegetable matter with shells including *Valvata piscinalis*; thickens northwards	0.08 to 2.13
Blue clay, yellow in places, with race; laminated; passes laterally into laminated clay with small flat calcareous and subglobose concretions. Dip to south and east; contains freshwater shells in the upper part, particularly *Valvata* and *Bithynia*	*c.* 3.40
Sandy gravel with much chalk thickness not stated	

A more complete section, incorporating Brown's (1852) section and further excavations of 1874, was described by Dalton (1880). The exact location of these sections is not known:

	Thickness m
Brown clay with boulders derived from Boulder Clay	0.9
White sand, shelly	0 to 0.9
Shell marl consisting of white homogeneous unctuous clay, shelly sand and ferruginous sand and clay; shells and bones are described by Brown (1843; 1852) from this horizon	1.5 to 1.8
Peat, contains freshwater shells; at the north of the brickfield it is 2.1 m thick, only 0.3 m in the north-west pit where it thins away southward, and only 0.15 m in the east pit	0.15 to 2.1
Brickearth; dark blue micaceous and calcareous clay with race nodules and freshwater shells; 4 m thick in the west, and 3.4 m in the east	3.4 to 4.0
Grey sandy gravel with chalk and a few fossils	*c.* 3.1
Brickearth; finely laminated bluish grey calcareous loam turning whitish when dry, and weathering at the surface to a rich brown loam; race nodules near the surface	5.5
Grey sandy gravel at the southernmost point probably Glacial Sand and Gravel)	no thickness given

The uppermost three beds are now regarded as Flandrian and not part of the Hoxnian succession (Turner, pers. comm.).

A tongue of Lacustrine Deposits extends up a shallow valley at the eastern edge of the outcrop [9355 2445] north-east of Copford Place. Whereas the general level of the surface of the Lacustrine Deposits rises from about 24 m beside the Roman River to 26 m, it reaches about 30 m OD at the head of the tongue. The thickness of the deposits hereabouts is not known but is likely to be less than 5 m; as a result only a small amount of compaction has occurred.

At Copford Place a trial pit [9320 2415] illustrates the nature of the marginal Lacustrine Deposits:

	Thickness m
Head	1.05
Lacustrine Deposits	
Smooth brown clay with irregular disturbed base	0.10
Yellow to buff and grey silt with variable clay content; abundant race at 0.50 m, scattered elsewhere	0.65
Lens of laminated silt and fine-grained sand	0.20
Lens of sand and fine gravel with rounded chalk and subangular flints < 2 cm diameter	0.03
Grey and orange mottled laminated silt; fining upwards beds; laminae < 1 mm thick	seen to 0.97

RAE

RIVER TERRACE DEPOSITS

River Terrace Deposits occur in the valleys of the Stour, Colne and Pant–Blackwater. Direct correlations between river systems is not possible in the district, but the terrace surfaces in each system fall into five altimetric groupings with respect to the present-day river flood-plains. Their general levels above the alluvium level are as follows:

Terrace 1	2 m
Terrace 2	3 to 9 m
Terrace 3	10 m in Colne valley; 15 m in Stour valley
Terrace 4	Blackwater valley only 11 m
Terrace 4–5	22 m

The highest terraces (4 and 4–5) have been distinguished only in the Blackwater valley near the southern margin of the district.

The numbering of the terraces is provisional because only parts of the river systems occur in this district and they cannot be related to the full river profile and, more particularly, to the river mouths. The Stour and Colne rivers in the eastern part of the district, however, lie almost at sea-level and it is assumed that their terraces are coeval. Development of the Pant–Blackwater terraces may have been affected by local factors of sediment supply, and the deposits are not directly correlatable with the Stour and Colne terraces. The Blackwater Terraces 1 and 2 described by Bristow (1985) in the adjacent Chelmsford district are not differentiated in this account. Terrace deposits crop out almost continuously on one or both sides of the Stour and also along the Colne upstream as far as Chappel [895 284]. In the Pant–Blackwater, terraces are only well developed south of Feering [860 205].

In many instances a planar terrace feature can be mapped, without a break of slope, from alluvium level up to heights consistent with Terrace 2 and, in the Stour valley to the immediate east of this district, to the height of Terrace 3. This apparent transition from one terrace surface to another illustrates the complexity of both the deposits and the associated features, and the problems of correlating terrace levels along a river system. In this district Terraces 1 and 2 crop out adjacent to alluvium and with one exception, in the Colne valley south of King's Farm [936 280], there is no London Clay bluff separating the deposits. London Clay invariably crops out below older terraces.

There is no strong lithological difference in the composition of the various terraces. At surface they give rise to soils containing mostly brown or patinated angular flints, but there is also a more silty soil derived from Head particularly near the back of the terrace feature. Some terrace features are blanketed by a layer of silt which represents river over-bank sediments. The gravel clasts consist of about 60 to 80 per cent subangular flints, along with subrounded flint, quartz, quartzite, sandstone, chert and other erratics derived from the Glacial Sand and Gravel. Rounded chalk pebbles are typically found at levels of more than 3 m below the surface and are particularly abundant, even at the surface, where the deposits lie on Chalk.

Gradings of samples of sub-alluvial gravels and terraces from each river system have similar overall characteristics, but are rather variable within each river system. The deposits are generally poorly sorted and the gravel content of individual beds within a terrace deposit varies from 20 to 80 per cent. There is no apparent trend in grading in any one terrace deposit; some are finer at the base whilst others fine upwards.

Terrace 1 and 2 deposits have been extensively excavated for aggregate in the district, but all the workings are now overgrown and degraded or backfilled, so that during the survey no fresh sections were visible. Such degraded exposures show rather poorly bedded deposits of rusty brown flint-rich sandy gravel, with some clayey horizons presumed to have been laid down in a braided river system.

The Blackwater terraces

Blackwater Terraces 4 and 5, which lie at up to 46 m OD, are only thin, as shown by pits less than 3 m deep formerly opened in the gravel deposits which overlie Boulder Clay.

Terrace 3 around Kelvedon is the most widespread of the Blackwater terraces. Its base is irregular and its back-feature lies at about 30 m OD but falls to about 15 m OD at Witham, 3 km south of the district. The precise age and origin of the terrace is not clear but it is probably the product of more than one aggradation phase, the earliest of which deposited the Glacial Sand and Gravel. Bristow (1985) regards the sands and gravels as outwash laid down in an irregular basin vacated by the melting Anglian ice-sheet. This is supported by the presence of Hoxnian deposits (see p. 40) overlying Terrace 3 at Rivenhall End, only 1.5 km south of the Braintree district (Turner *in* Rose and Turner, 1973; Bristow, 1985).

Blackwater Terraces 1 and 2 are preserved locally, mainly at Feering [867 205], between Stisted [800 240] and Bradwell [834 223], and near Bocking Church Street [762 251]. Terrace 2 occurs at 52 m OD in the upstream outcrops, falling to 26 m OD at Feering where it is grouped with Terrace 1.

The Stour and Colne terraces

The highest terrace feature elevation in the Stour valley is found in a broad embayment to the east of Mount Bures [91 32]. The terrace back-feature lies at between 33 and 46 m OD and, as there is no comparable terrace elsewhere in the district, it has been designated as 'Terrace undifferentiated'. Information is limited, but the deposits are known to vary in thickness from less than 2 m to more than 3.5 m, and to include sand and gravel and greenish grey lacustrine laminated clays.

A small tract of Terrace 3 at 32 m OD in the Stour valley lies about 15 m above the flood plain south of Elm's Farm [914 320] in the embayment described above, and a further small area occurs at Lower Dairy Farm at 29 m OD. The thickness of the terrace is not known but field relations suggest that less than 4 m are present.

Terrace 3 in the Colne valley lies at about 11 m above Alluvium, and its back-feature falls from 32 m OD east of Bacon's Farm [901 271] to 18 m OD near Lexden Lodge [977 264]. In most cases the front edge of the terrace forms a well marked break of slope on to underlying London Clay but locally, for example north of Chitts Hills [958 261], the surface grades gently down on to lower terrace levels. The only borehole to penetrate it [9572 2614] proved 2.7 m of clayey gravel, whilst degraded excavations near Lexden Lodge [977 262] suggest that about 3 to 4 m of gravel were extracted. A former brick-pit at Lexden, whose exact site is uncertain, but which has yielded a large fauna and flora, is

recorded as being 11 m above river level, and is thus consistent with the Colne Terrace 3. The site was originally described by Fisher (1863) as 'on a plateau upon the south side of the valley of the Colne about a mile west of Colchester'. The pit has since been backfilled, and the approximate grid reference for it [978 253] given in Shotton, Sutcliffe and West (1962) is in the Colchester Castle grounds where an old brickworks was dug into London Clay (Colchester Nat. Hist. Museum, *pers. comm*); the grid reference, therefore, appears to be incorrect. A more likely site is 1.1 km east of Maltings Farm [9800 2545], but this was ploughed over even before Dalton's survey (1880). According to Fisher (1863) the terrace deposits consist of brickearth overlying gravel in which is a channel containing peat and organic clay. A large flora and fauna collected by him and now deposited in the British Museum (Natural History) includes beetles and teeth of *Elephas primigenius* and *Rhinoceras leptorhinus*. A reappraisal of the fossils and of pollen extracted both from the peat and from the mammal teeth enabled Shotton, Sutcliffe and West (1962) to state that a climate not unlike that in modern Essex is likely to have prevailed during deposition of the peat, although they came to no conclusion about its age.

The second terrace is generally the better developed of the two lower ones; in the Stour valley it falls from 28 to 20 m OD whilst in the Colne valley it is at 47 m OD just south of Sible Hedingham [787 335] dropping to about 10 m OD at the eastern margin of the district, a fall of 37 m in 33 km. The feature associated with Terrace 1 is not always clear, and in many places no front feature is present, terrace gravels being overlapped by Alluvium. Weirs and river diversions made for water-powered mills and, more recently, for flood prevention have considerably modified the valley topography so that differentiation of these low terrace deposits is difficult.

Boreholes put down on Terraces 1 and 2 and on alluvium have bottomed gravels at very variable depths. In the upper reaches of the Colne valley, in particular, much of this gravel is thought to be glacial in origin (see p. 30). The main feature of the Stour is the relatively thick gravel sequence, up to 18 m beneath Terrace 1 and Alluvium at Bures. Elsewhere in the Stour valley the base of the terrace gravels generally lies less than 10 m below alluvium.

The Colne valley profile is longer and provides a more complete picture although borehole information is sparse. Up to 20.9 m of gravels have been proved near Earls Colne Priory in one borehole [8658 2857], and similar sequences have been penetrated in a number of others. From the available information it seems likely that thick gravels, partly glacial in origin, are present beneath the valley, thinning gradually downstream. RAE

Details

Blackwater valley

There is only one small spread of Terrace 5 [878 184]; it consists predominantly of sandy clay, with minor amounts of sand and gravel. It is probable that the spread of sand and gravel east of Yewtree Farm [887 188] mostly belongs to Terrace 5. In this tract the surface indications are of a very gravelly deposit.

West of Yewtree Farm, the gravel spread assigned to the combined terraces 4 and 5 declines northwards from 38 m to 30 m OD. It is dominantly gravelly, but it is not thick; small pits [8857 1924; 8863 1918; 8865 1908] formerly opened in the sand and gravel, terminated at shallow depth in Boulder Clay. Dissected remnants of a presumably once more widespread Terrace 4 lie in the vicinity of Park Farm [878 188]. The deposit appears to consist of medium-grained sand and clayey sand and gravel. A borehole [8699 1832] proved only 1 m of sand and gravel above Boulder Clay.   CRB

An outcrop [823 225] of Terrace 3 deposits is present near Stockstreet. Sandy gravels were noted near the surface of this deposit which rises to about 10 m above the alluvium. One borehole [8756 2043] near Feering proved 4.9 m of terrace gravels beneath 1.2 m of Head and resting on Boulder Clay. Clayey sands are present at the surface which lies at levels up to 13 m above the floodplain.   RDL

Near Gore Pit [871 197] the gravel proved in a well [8721 1968] is 6 m thick, and rests on Boulder Clay (Whitaker and Thresh, 1916, p.155). The base of the terrace deposits lies at 26.8 m OD. The gravel thins northwards, and Boulder Clay is reached in the bottom of pits [8720 1991; 8715 2000] in the Little London area. During the excavation for the northern part of the Kelvedon bypass, Terrace 3 was seen to thin northwards, from about 4 m at a point [8776 1959] north-east of Threshelfords Farm, to less than 1 m some 500 m farther north [c.877 200]. The thickest sand and gravel proved along this part of the bypass was 7.55 m [8766 1956]. Another borehole [8710 1926], some 250 m east-south-east of Threshelfords Farm, proved the Terrace 3 deposits to be tripartite, with an upper unit of sand and gravel, 1.7 m thick, overlying laminated clays (see p. $\overline{40}$), 2.3 m thick, which in turn overlay 6.4 m of sandy gravel, above Boulder Clay.

On the west side of the River Blackwater the surface indications are of a very gravelly deposit. A well [8647 1916] at the now demolished Fuller's Brewery, Kelvedon, proved 0.6 m of made ground, above 9.5 m of sand and gravel, overlying Boulder Clay at 15.5 m OD. South of the brewery, a borehole [8639 1867] passed through 8.2 m of top soil and clayey sand and gravel of Terrace 3 before entering Boulder Clay at 17.4 m OD. Another borehole [8556 1829] proved 14.9 m of top soil and lacustrine deposits (see p. $\overline{40}$), overlay more than 9.5 m of pebbly sand; the base of the terrace lies below 1.2 m OD.   CRB

A tooth of 'Elephas primigenius' was collected in a pit dug in Terrace 2 deposits south of the 'church at Bocking'; it exposed 5.2 m of sandy gravel (Whitaker and others, 1878, p.72). The exact location of this section is uncertain, but it may have been at King's Bridge [7590 2545]. Deposits comparable to those at Sheering Hall (see below) are present below Bocking Churchstreet where the terrace gravels lie at up to 12 m above floodplain level, and have been worked locally for gravel in the past. Patches of Terrace 2 deposits are present from below Stisted Hall to Stockstreet. They are predominantly gravelly in character and have a surface level which typically ranges from 2.5 to 3.7 m above the floodplain, and locally extends up to 6 m. Disused small pits [803 237] are present near Shelborn Bridge. Near Feering the outcrop [868 206] of the combined terraces 1 and 2 comprises clayey silts and sands about 1 m thick, overlying gravels of unknown thickness.

Near Sheering Hall an area of sandy gravels with a marked back-feature about 9 m above the floodplain level has been ascribed to terrace 1 and 2. Locally the gravels are thin and impersistent, and in the western part an extensive clay wash overlies them.   RDL

Colne valley

Small patches of Terrace 3 occur at up to 13 m above alluvium level east of Bacon's Farm [8982 2715]. The soil is very gravelly, and a small pit [9023 2726] exposed 2.1 m of sand and medium-grade flint gravel. Further small tracts of Terrace 3 are to be found north of Chitts Hills [956 262], between 15 and 25 m OD. Their thickness is probably not greater than 3 m; one borehole [9572 2614], proved only 2.7 m of 'brown clay and gravel' resting on London Clay at 18.3 m OD. A well-defined terrace at Lexden Lodge [977 263] has been worked for gravel, and subangular to rounded fine to coarse gravel is abundant in the soil around the site of old pits [9735 2596]; the gravel consists mainly of brown coated flints with a few quartzite pebbles. On the southern bank of the River Colne, 700 m south of Lexden Lodge, Terrace 3 forms a flattish feature at about 16 m OD, some 8 m above alluvium level, and is characterised by numerous brown-coated flints in the soil.   RAE

In the Hedinghams area Terrace 1 forms good features with a general surface level of about 1.2 to 1.5 m above the floodplain. At the surface silty loams are general, although at Pool Street [766 369] a gravelly soil is common. At Halstead the surfaces of the terrace deposits locally reach 2.4 m, and generally lie at 1.5 to 1.8 m above the floodplain. The deposits appear to be dominantly silty at the surface although one tract is largely obscured beneath the urban area.

Near Earls Colne there are extensive outcrops of Terrace 1–2. These deposits, which have a variable composition at the surface, appear from borehole evidence to be between 5 and 7 m thick although thicker sequences have been recorded, in one example associated with buried channel deposits of probably glacial origin (see p. 30). These terrace deposits have levels of up to 6 m above the floodplain. To the east of Earls Colne [880 284], sandy Terrace 1 deposits have been worked to a depth of 3.7 m. At Chappel the Terrace 1 deposits have clayey loam with scattered pebbles at the surface.

Near the Hedinghams the Terrace 2 deposits comprise gravelly sandy loams at the surface, which is 2.4 to 3.7 m above the floodplain. North-west of Halstead, one outcrop [805 316] has a sandy surface. The Terrace 2 deposits near Chappel comprise fine- to coarse-grained sands and sandy gravels, and are present at levels up to about 4.5 m above the floodplain.   RDL

Between Wakes Colne [896 284] and Fordstreet [921 270] terrace features are well developed: Terrace 1 forms well defined, but small, tracts some 1 to 2 m above alluvium level [9070 2755; 912 270]. Terrace 2 occurs over large areas on the south side of the river 500 m south-east of Broom House [902 275]. It is veneered with a sandy soil and rises to about 5 m above alluvium level.

Undifferentiated Terrace 1–2 deposits [917 272], immediately west of Fordstreet, lie at about c.16 to 22 m OD, and were extensively worked for gravel during both World Wars. No exposures in the main worked area were seen during the present survey, but in a small pit [9165 2735] the upper 2 m of the terrace deposits consisted of orange-brown clayey sandy gravel comprising mainly subangular and nodular medium- to coarse-grade flint pebbles; at least 5 m of terrace deposits have been worked in the general area.

From Fordstreet to Cook's Mill [9480 2705] terraces form an almost continuous tract along the northern bank of the River Colne, where small areas of Terrace 1 present a very low feature less than 1 m above alluvium level near Mill House [9277 2720] and west of Cook's Hall [956 274]. Terrace 1–2 occurs east of Mill House and is undifferentiated because there is no change of gradient in the slope from alluvium level to above 22 m OD. There is a small gravel pit [9277 2730] in the tract at about 15 m OD. It is likely that these undifferentiated deposits have an uneven base since London Clay was augered within the outcrops [9367 2767].

Terrace 2 crops out east of Wash Farm [922 273], and south of Kings Farm [942 278; 945 278] forms strong features rising to above 22 m OD. The sand and gravel is mainly veneered by brown sandy silt although the soil tends to become increasingly gravelly towards the front feature. The Head mapped south of Cook's Hall may overlie terrace deposits although the features are modified by soil-wash and the slope is gently concave. On the south side of the River

Colne between Fordstreet and Cook's Mill, terraces are more restricted in extent than on the north side, but they form narrow features less than 50 m wide [9262 2712; 9342 2739]. The surface of Terrace 2 near Cook's Mill [945 270] rises 4 to 5 m from about 12 m OD at its front edge and the rough topography of the ground at the back of this terrace is probably the site of a degraded gravel pit.

Extensive tracts of Terrace 2 deposits occur at the confluence of Bourne Brook and the River Colne near Bourne Farm [9524 2659] and beside the nearby railway embankment. A large borrow pit [964 263] worked from 1972 to 1975 is reported to have yielded 'hoggin' and gravel for use in construction of the Colchester by-pass; 3 to 4 m of terrace deposits were removed but no exposures were visible during the present survey. Terrace 1–2 deposits north-west of Seven Arches Farm [9635 2595] were also worked and backfilled from 1972 to 1975. Gravel, clayey in places, was dug out to a depth of 4 m in places. In this vicinity the terrace feature rises from river level at 12 m OD to about 16 m OD.

Site investigation boreholes along the line of the Colchester by-pass and its slip road [966 257] provide the only indication of the total thickness of the River Colne terraces between Earls Colne and the eastern boundary of the district; the maximum proved thickness is in a borehole [9660 2577] where 14.8 m of sandy gravel with clayey beds lay on London Clay at about 4 m below OD. Another borehole [9655 2573], near Seven Arches Farm, penetrated 10.3 m of terrace sand and gravel.   RAE

Stour valley

A trial pit put down on undifferentiated terrace deposits, 100 m north-west of Elm's Farm [9142 3209], proved the following sequence:

	Thickness
	m
Top soil	0.30
Clay, brown, and pale grey mottled, variably silty, with rounded flints up to 3 cm and subordinate angular flints and rounded quartzite pebbles; gradational base	1.00
Clay, slightly silty, stiff greenish grey and brown mottled, reddish brown in places; increasing mottling towards the top; silt content increases downwards, with coarse silt at base; irregular distribution of calcareous nodules throughout; sharp base	1.60
Gravel, of rounded, angular and subangular flints in a coarse sand matrix; poorly sorted and with clay lenses and some clay in matrix; a few large flint cobbles up to 10 cm diameter	seen to 0.40

Terrace deposits 1 and 2 north of Bures are generally from 6 to 7.5 m thick, and reach a maximum elevation of 8 m above alluvium level. Predominantly flint gravels of Terrace 1 and 2 have been proved beneath Alluvium in boreholes [8857 3775; 8942 3677]. The gravel in the former borehole contained some vein-quartz pebbles, large flint cobbles and chalk fragments; in the latter hole the gravel was apparently chalk-free.

Near Bures the outcrop is much narrower, and the gravels are appreciably thicker than upstream, and reach 17.7 m in a well [9076 3400]. Mixed clay and gravel to a depth of 9.1 m were recorded in two other wells in Bures [9046 3397; 9052 3404].

East of Bures many of the terrace surfaces which slope gently riverwards are veneered in brown silt with scattered gravel, usually less than 1.5 m thick; boreholes prove the terrace deposits here are in total 5 to 7 m thick; exceptionally, at Staunch Farm [9194 3282], a borehole proved 8.9 m of gravel overlying Lower London Tertiaries. The terrace gravels are mostly of flint, but vein-quartz, quartzite pebbles and hard chalk pellets were recorded in most

boreholes (Hopson, 1981). Terrace 2 south of Bowdens [9405 3340] has a gravelly soil near its boundary with the alluvium, and gravel has been extracted from a small pit [9497 3246].

Terrace 1–2 to the north-east of Wissington [947 332] has an even surface gradient from its back-feature to the alluvium, where the front of the terrace is marked by a small feature 1.5 to 2 m high. Brown silt about 1 m thick, overlying gravel, was augered in some ditches crossing the terrace, but more than 2 m of silt are present locally, particularly towards the back of the terrace. In a borehole [9580 3353] near Wissington, 1.1 m of pale yellow-brown and grey mottled sandy silty clay with some flint pebbles was recorded, overlying 6.6 m of gravel, clayey at some levels, with a dark grey silt band 2.0 m from the base.   RAE

HEAD AND HEAD GRAVEL

Certain heterogeneous deposits which do not fall into a specific genetic division have been grouped under the term Head. They have in all cases been derived from local parent material, probably initially by solifluction processes in periglacial conditions, but continuing at the present day as soil-creep and minor mass-movements. Typically, therefore, the lithologies encountered reflect the source material upslope, although on plateaux where mass movement has been only limited, they may represent the degraded and cryoturbated relics of thin drift deposits. Spring-line deposits also augment slope material with fine-grained sediment derived from the arenaceous drift deposits.

In general, the Head deposits are poorly sorted clayey silts and sands, generally brown, with a subordinate gravel content which is normally concentrated at the base; Head Gravel is a poorly sorted clayey gravel. Head occurs widely in all the river valleys and many of the tributaries in this district, commonly obscuring the contact between the Kesgrave Sands and Gravels and the underlying London Clay. It has been seen to reach 3 m in thickness, and only deposits thicker than 1.5 m have been depicted on the geological map.

Head Gravel has been distinguished around Wethersfield where it is probably less than 2 m thick and is derived from Boulder Clay and Glacial Sand and Gravel by solifluction in the Pant valley; elsewhere, it has not been practical to map Head Gravel deposits separately.

The age of formation of the Head is difficult to establish, and most probably it has been accumulating since the deposition of the glacial sediments, particularly during cold periods. Polycyclic solifluction deposits have been recognised, for example in the Yeldham area and in the Belchamp Brook valley, that may represent solifluction lobes accumulated in successive cold periods. In these deposits, coarse clastic material tends to be concentrated towards the sole of each lobe, presumably due to differential settlement during emplacement.

Details

The Pant – Blackwater valley

Bennett (*in* Whitaker and others, 1878, p.66) noted that on the western side of the lane north of Petches Bridge, west of Wethersfield, a pit [*c.*6995 3129] showed 2.4 m of rather coarse subangular gravel with a bed of sand near the bottom. In this sand he found bone material, probably of elephant. Large bones were also recorded at Daw Street to the north-west.

In the Wethersfield area, the Head comprises flinty sandy clays with pockets of peaty material, especially near the spring-line at the base of the Kesgrave Sands and Gravels where these overlie London Clay. The Head Gravels in this area consist of clayey gravels with brown patinated flints.

The Colne valley

In the Colne valley near the Hedinghams, the Head consists of a brown sandy clay with much angular flint material and, where observed in a railway cutting [7730 3621] north of Sible Hedingham, it is clearly divisible into two units representing two different solifluction lobes.

The Head Gravel near Kirby Hall [779 372; 773 368] consists of clayey gravels with brown patinated angular flints; a trial pit [7793 3731] proved:

	Thickness m
Gravelly top soil	0.30
Medium to coarse sand passing downwards into brown, clayey, fine to medium gravel with rounded, angular and nodular flints; distinct convoluted clay partings; sharp irregular base	1.10 to 1.50
Chalky Boulder Clay	seen to 0.40

The Stour valley

In the Stour valley two types of Head have been recognised: a common, brown, flinty, sandy clay or silt, and a more local clayey flint gravel, as for example at Kedington Hill and in the Wiggery valley. A trial-pit [8240 3814] north of Wickham St Paul's showed evidence of three phases of solifluction (Beds A, B and C, below):

	Thickness m
Top soil	0.3
Bed A: Sand, medium- to coarse-grained with fine to coarse gravel mainly 2 to 3 cm diameter consisting of angular to well rounded flint and rounded quartz and quartzite clasts; clayey in places; dirty orange brown; sharp irregular base	0.70 to 1.00
Bed B: Sand, fine- to medium-grained, and silty clay, brown to orange-brown, convoluted; the clay contains pellets (c.5 mm diameter) of iron-cemented sand and small carbonaceous flecks; at the base is a gravel up to 20 cm thick containing rounded black flints and quartzite pebbles; a clay band 2 cm thick forms the sole of this solifluction lobe	0.75 to 1.45
Bed C: Silt and fine-grained sand with clay bands; buff to brown with brown to red-brown and grey mottling; convoluted; sharp sub-horizontal base (derived from beds below)	0 to 0.75
Woolwich and Reading Beds	seen to 1.00

RDL

ALLUVIUM

Alluvium occurs in all the major river valleys of the district and in many of their tributaries. It forms flat flood-plains which are subject to flooding during high river discharge, even in the Stour and Colne valleys where weirs have been constructed. The boundaries of Alluvium are drawn at the breaks of slope against the adjacent formations, usually terraces or Head, a line commonly coinciding with drainage ditches. Flood-meadows, alder carrs and willow plantations characterise the alluvial tracts.

The widest tracts are in the Stour valley where, apart from a constriction near Bures where it is less than 100 m wide, the flood-plain is between 200 and 800 m wide. Elsewhere in the district alluvial tracts are rarely more than 250 m wide and dwindle in the upstream parts of the rivers Colne, Brain and Pods Brook to less than 50 m.

The only exposures of Alluvium are in fresh ditch-sections and in river banks on the outsides of meanders where it consists of very soft, pale grey and brown silt to silty clay usually containing organic detritus. Near the margins it may contain gravel beds which are probably contiguous with Terrace 1 deposits. In the Stour valley, borings have proved that there is generally less than 2 m of Alluvium underlain by gravels (see p. 49). Exceptionally a greater thickness is present, as in a borehole [8942 3677] near Daw's Hall where 6.5 m of soft, mottled brown and grey, silty clay with organic debris and shells were proved.

Borings in the Colne valley have in general proved rather thicker alluvium sequences than in the Stour; these consist of up to 5 m of peat and soft grey and brown silty clay. Excavations [9291 2712] for a bridge-crossing at Mill House, Fordham, during this survey exposed the base of the Alluvium resting on gravels. Samples of wood and organic detritus collected from immediately above the base have yielded a radiocarbon date of $12\,075 \pm 65$ years BP.

The upstream part of the Colne has some anomalously thick sequences of Alluvium; a boring [7757 3618] put down in the valley north of Castle Hedingham proved silts with organic detritus and thin-shelled bivalves to 17.8 m below ground level. The relationship of these deposits to the thick gravel sequences proved in nearby boreholes is not known, and there is no dating or palaeontological identification to assist interpretation. It seems likely, however, that at least the lower parts of the deposit may be the record of river backwater or abandoned meander sediments formed during a phase of gravel aggradation in the Colne valley.

The River Blackwater alluvium reaches 9.5 m thick at Coggeshall, but it is not possible to determine its age or relationship to the sub-alluvial gravels. RAE

PEAT AND CALCAREOUS TUFA

Calcareous Tufa is formed locally at springs emanating from the Kesgrave Sands and Gravels where the water is lime-rich. Peat is also found locally where particularly wet conditions or poor drainage have persisted for some time; peat deposits which occur within Alluvium are discussed above.

Details

Three small areas of peat [7525 2040; 7560 2035; 7610 2040] in association with springs issuing from the base of the Kesgrave Sands and Gravels were mapped by Dr Allender in a small valley west of Black Notley Hall. At the last locality the peat is associated with calcareous tufa.

Two small patches of Calcareous Tufa have been noted in the Brain valley. The larger of the two spreads measures 70 × 50 m, and is forming at springs issuing from the junction of the Kesgrave Sands and Gravels and London Clay [7767 1901]. The other, much smaller, patch measuring some 30 × 10 m [7829 1946] is situated on the opposite side of the valley and lies close to the same boundary.

Calcareous Tufa is associated with peat [7610 2040] in the small valley west of Black Notley Hall. CRB

LANDSLIPS

There are a few small areas of landslipped London Clay on the south side of the River Stour near Wormingford [9265 3256; 9300 3272; 9610 3275] on the south side of the River Colne at Chappel [8870 2790] and near Eight Ash Green [9340 2725; 9465 2668]. The instability of the slopes is caused by oversteepening by river undercutting. Failure of the London Clay, which occurs on slopes greater than 7°,

has occurred because of a combination of the comparatively steep slope and the effect of groundwater emanating from springs at the Kesgrave Sands and Gravels – London Clay junction. The groundwater gives rise to a raised moisture content and high pore pressures in the London Clay, with a consequent reduction in its strength leading to its eventual failure.

The landslipped ground is characteristically uneven, and the mass-movement is in most cases in the form of mudslides affecting a slab of ground no more than 2 to 3 m thick. In the case of the landslip at Wormingford [9265 3256], there is borehole evidence of a large scale mass-movement affecting several metres of strata. The Wormingford Borehole (Ellison, 1976 and Appendix 1), put down on the eastern edge of this landslip, proved 4.5 m of mottled orange-brown disturbed London Clay underlain by 0.2 m of a soft wet peaty clay before undisturbed London Clay was reached. RAE

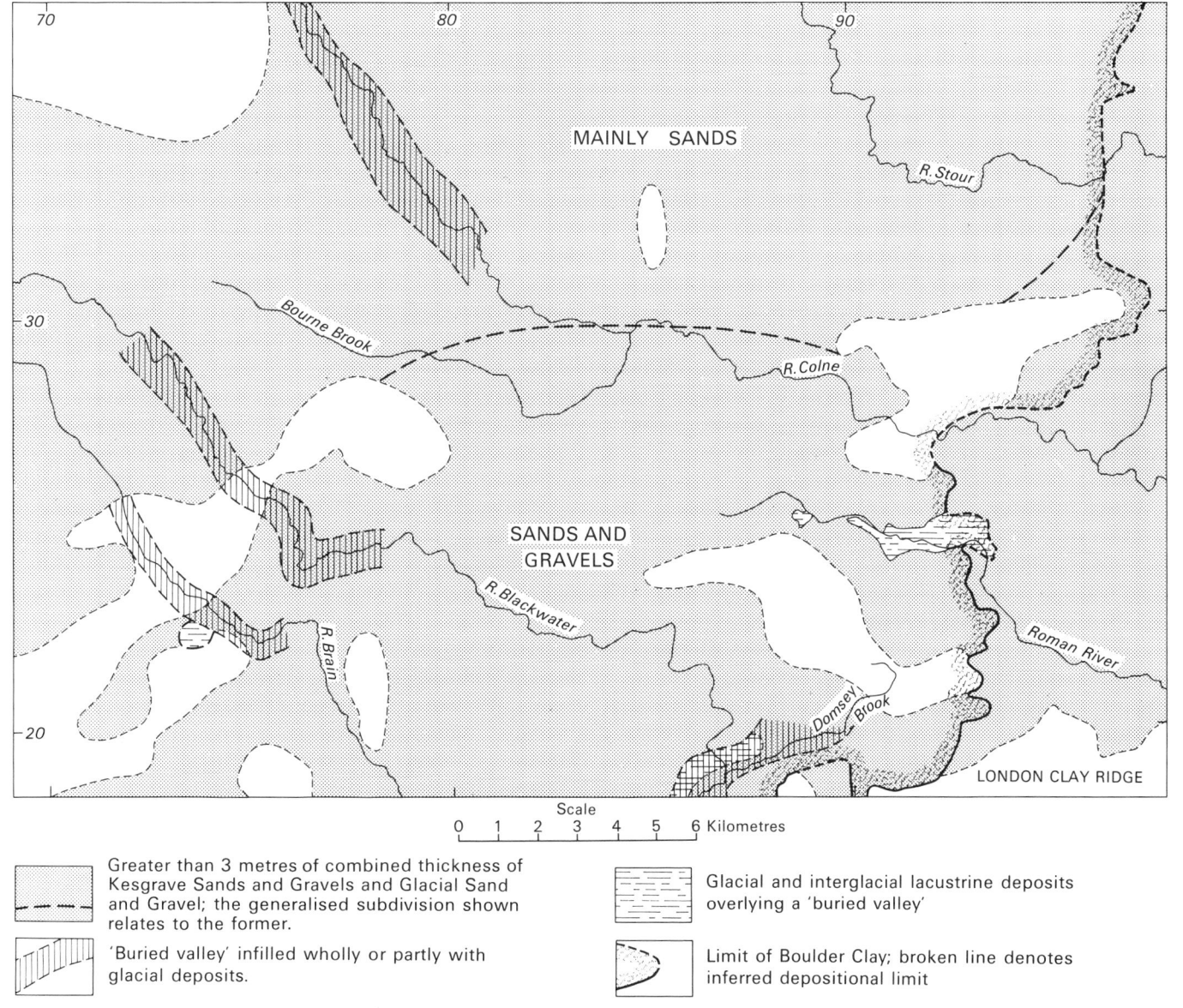

Figure 17   The distribution of selected drift deposits: the effects of erosion in the valleys are omitted

CHAPTER 7

Economic products

SAND AND GRAVEL

Following a feasibility study, which was conducted partly at Martlesham Heath (Suffolk) and partly in the Terling and Maldon areas of Essex during 1966 and 1967, the Industrial Minerals Assessment Unit of the Geological Survey has been concerned with the assessment of resources of potentially workable sand and gravel in many parts of the United Kingdom. This account outlines the results of surveys undertaken between 1967 and 1978, which form the basis of ten reports (see Figure 18) covering 97 per cent of the Braintree district (about 550 km²) in which there were found to be extensive resources of sand and gravel in the Crag, Kesgrave Sands and Gravels, Glacial Sand and Gravel (Figure 17) and River Terrace Deposits. Excluding the urban areas, 433 km²

are mineral-bearing and it is estimated that total resources amount to approximately 3700 million cubic metres (5600 m tonnes), of which over 96 per cent, 3585 million cubic metres (5400 m tonnes), are found in the Crag, Kesgrave Sands and Gravels and Glacial Sand and Gravel, with the remainder in the River Terrace Deposits.

For the purpose of the surveys a sand and gravel resource was regarded as potentially workable if a deposit met the following four arbitrary criteria:

1 The deposit should average at least 1 m in thickness
2 The ratio of overburden to sand and gravel should not exceed 3:1
3 The proportion of fines (particles less than 0.063 mm diameter) should not exceed 40 per cent (see Figure 19)
4 The deposit should be within 25 m of the surface

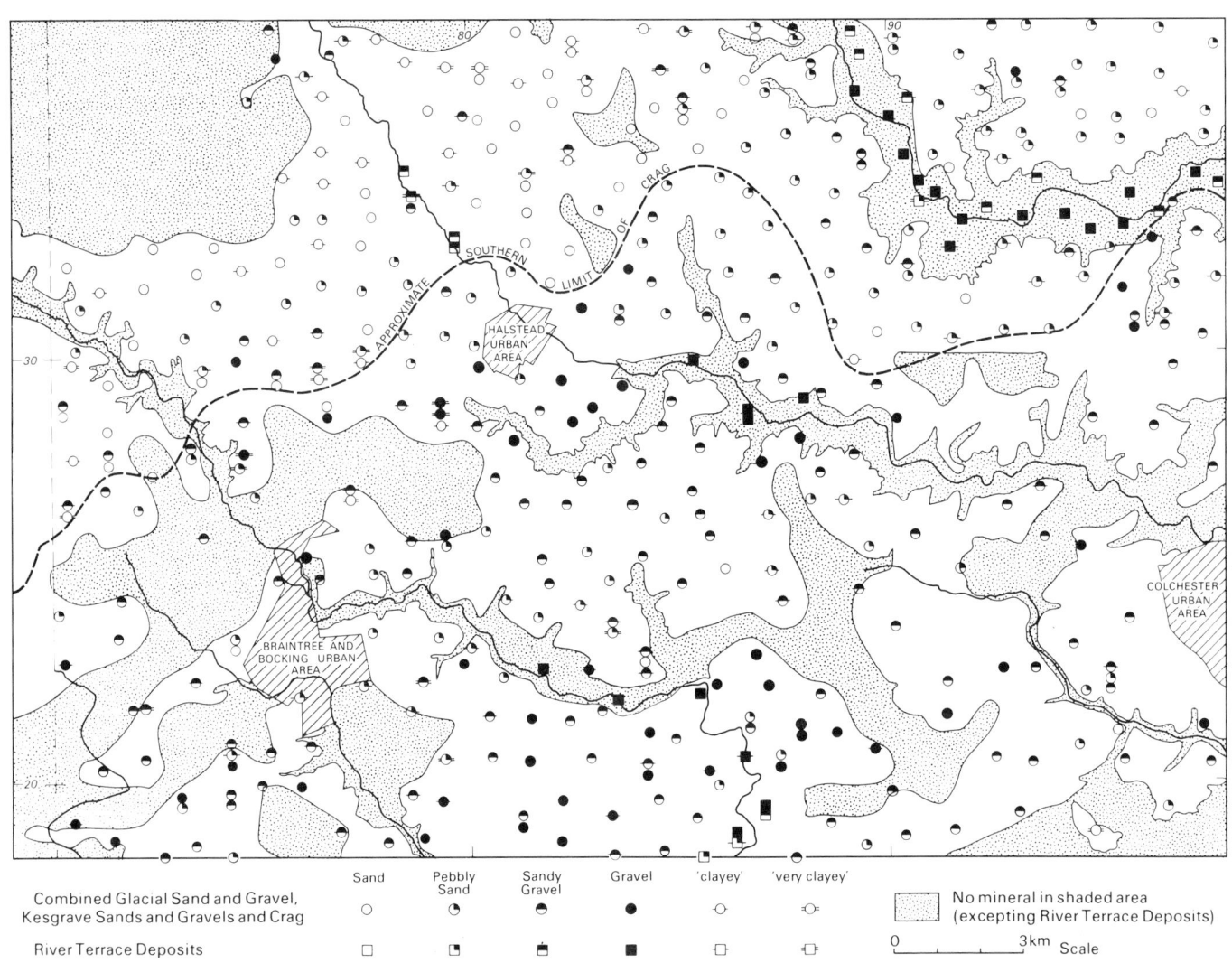

Figure 18 Classification of Mineral Assessment boreholes

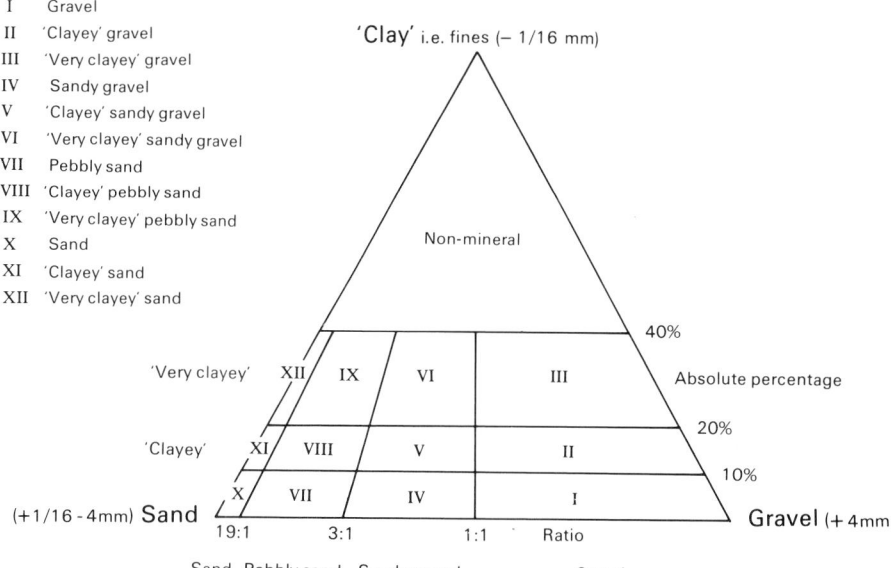

I   Gravel
II   'Clayey' gravel
III   'Very clayey' gravel
IV   Sandy gravel
V   'Clayey' sandy gravel
VI   'Very clayey' sandy gravel
VII   Pebbly sand
VIII 'Clayey' pebbly sand
IX   'Very clayey' pebbly sand
X   Sand
XI   'Clayey' sand
XII   'Very clayey' sand

Figure 19 Diagram to show the descriptive categories used in the classification of sand and gravel

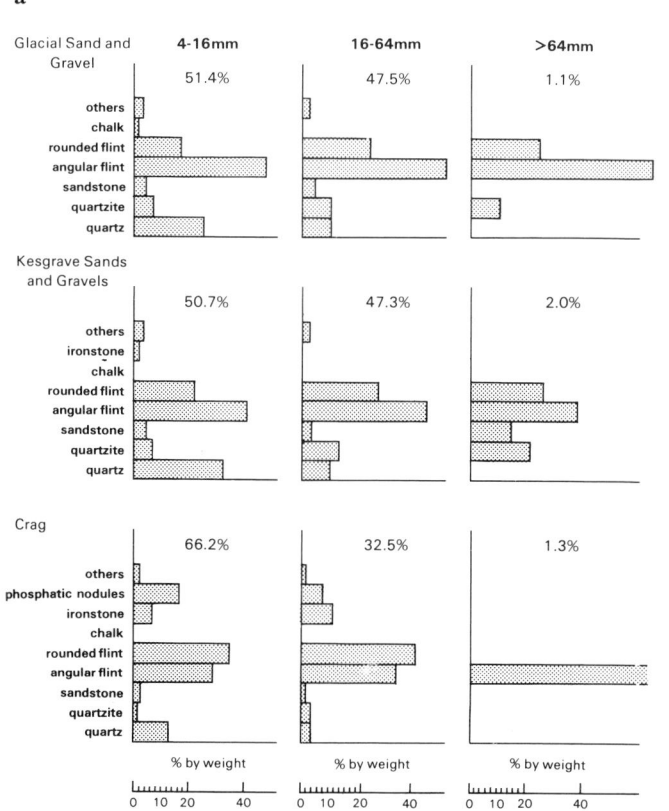

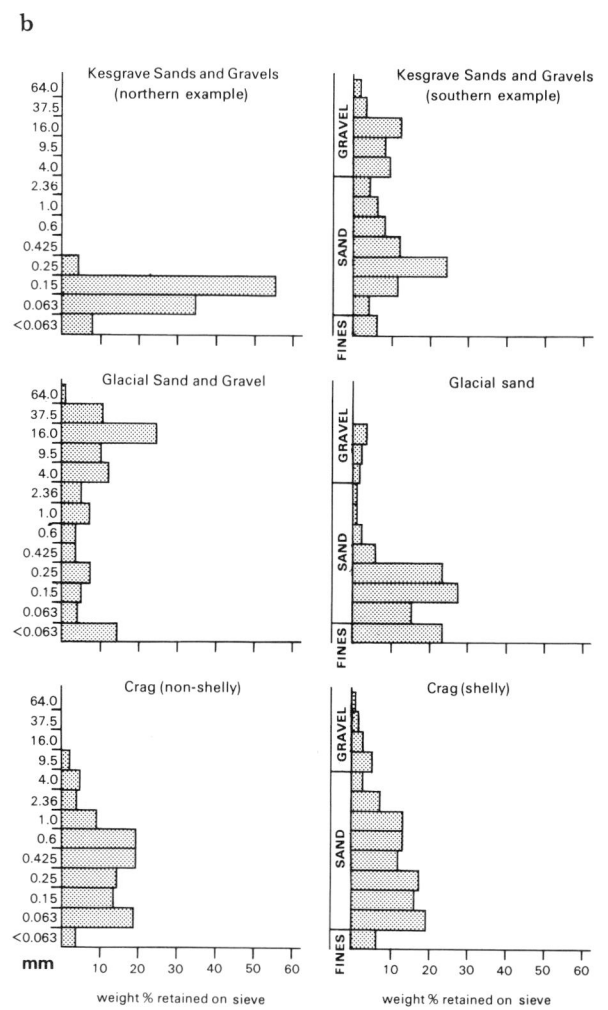

Figure 20 **a** Distribution by weight of clast type in different size ranges (of the Glacial Sand and Gravel, Kesgrave Sands and Gravels and Crag)
b Grain size histograms (of the Glacial Sand and Gravel, Kesgrave Sands and Gravels and Crag)

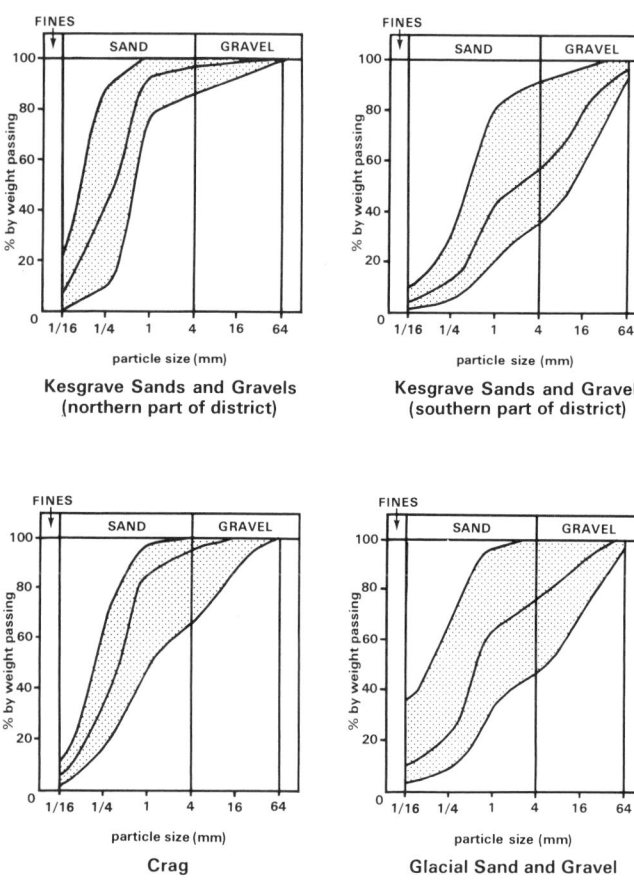

Figure 21 Representative grading curves showing the mean and range within which all samples fall

Crag

The Crag is described mainly as a shelly sand. In the area of Sheets TL 73, 83 and 93 it has been further distinguished by its characteristic grading curve and pebble content (see Figures 20 and 21). Using these criteria, boreholes in the assessment surveys have proved considerably more Crag deposits than are shown on the 1:50 000 geological map and its extent is illustrated on Figure 18. It is predominantly sand or pebbly sand, commonly more gravelly at the base; a basal shelly bed occurs locally. The maximum proved thickness is 6.5 m in Borehole 93 NE 32 [9547 3525], thinning towards the west and south. The mean grading of the Crag is 6 per cent fines, 89 per cent sand and 5 per cent gravel. The gravel fraction is composed predominantly of well-rounded and angular flint, with tabular ironstone and well-rounded quartz and phosphatic nodules, and a few quartzites, sandstones and igneous and metamorphic rocks. (see Figure 20).

Kesgrave Sands and Gravels

This formation forms the greatest resource in the district ranging from fine sand to gravel (see Figure 21). The gravel and sandy gravel categories have generally less than 10 per cent fines, whereas the pebbly sand and sand are commonly 'clayey' and less commonly 'very clayey' because of discrete laminae of grey clay which occur sporadically throughout the sequence.

Table 3 List of Sand and Gravel operations

	Grid ref.	Geological formation worked
Working pits		
Foxborough Hill, Halstead	795 324	Kesgrave Sands and Gravels
Bradwell	817 216	Kesgrave Sands and Gravels
Ferriers Farm, Bures	895 342	Glacial Sand and Gravel, Kesgrave Sands and Gravels and Crag
Brake's Farm, Birch	927 200	Glacial Sand and Gravel and Kesgrave Sands and Gravels
Bellhouse Farm, Stanway	947 223	Glacial Sand and Gravel and Kesgrave Sands and Gravels
Warren Lane, Stanway	948 229	Glacial Sand and Gravel and Kesgrave Sands and Gravels
Warren Lane, Stanway	950 225	Glacial Sand and Gravel and Kesgrave Sands and Gravels
Disused pits		
Shalford	722 286	Kesgrave Sands and Gravels
Beazley End, Shalford	735 289	Kesgrave Sands and Gravels
Great Codham Hall, Shalford	738 283	Kesgrave Sands and Gravels
Rayne Road, Braintree	746 228	Kesgrave Sands and Gravels
Straits Mill, Bocking	770 247	Kesgrave Sands and Gravels
Well Farm, Gosfield	786 296	Kesgrave Sands and Gravels
Bradwell	803 230	Kesgrave Sands and Gravels
Silver End	810 188	Kesgrave Sands and Gravels
Fabians Farm, Coggeshall	855 233	Kesgrave Sands and Gravels
Alphamstone	870 355	Kesgrave Sands and Gravels
Colne valley	875 285	River Terrace Deposits
Wash Farm, Fordstreet	917 273	River Terrace Deposits
Shrub End	968 232	Glacial Sand and Gravel and Kesgrave Sands and Gravels

In general the resource classed as sand and pebbly sand lies around Halstead, Nayland and Great Maplestead where up to 15 m occur. Beds falling within the gravel and sandy gravel category occur mainly in the south of the district. These two lithological types are interbedded in an area from Pebmarsh to Braintree. The mean grading of the Kesgrave Sands and Gravels is 7 per cent fines, 72 per cent sand and 21 per cent gravel; the gradings of the two end members are shown in Figure 21. The gravel fraction is composed predominantly of angular and rounded flint, rounded quartz and quartzite, together with some sandstone and minor amounts of ironstone, limestone, and igneous and metamorphic rocks. (see Figure 20).

Glacial Sand and Gravel

Much of this sand and gravel has a very variable fines content, and samples commonly fall in the 'clayey' and 'very clayey' descriptive categories. It ranges from a sandy gravel to beds of pebbly sandy clay, classed as 'non-mineral', which in many places are sufficiently thick to be designated as 'waste', a term which denotes material occurring within or below a sand and gravel resource but considered not to be potentially workable.

Glacial Sand and Gravel has a mean grading of 13 per cent fines, 65 per cent sand and 22 per cent gravel (see Figure 21); the gravel fraction consists mainly of angular flint, with well-rounded quartz and flint, and minor amounts of quartzite, sandstone and igneous and metamorphic rocks (see Figure 20).

River Terrace Deposits

Potentially workable sand and gravel occurs beneath Alluvium and as terrace deposits in the Stour, Colne and Blackwater valleys. The mean grading of River Terrace Deposits is 6 per cent fines, 47 per cent sand and 47 per cent gravel; the gravel fraction commonly contains flint cobbles (greater than 64 mm diameter) but consists mostly of angular flint with some subrounded to rounded quartz and quartzite, chalk and igneous and metamorphic rocks (see p. 48). Chalk accounts for 1 per cent by weight generally, although up to 10 per cent is locally present where Terrace Deposits rest directly on Chalk bedrock in the upper reaches of the Colne valley and in the Stour valley. PMH

REFERENCES

ALLENDER, R. and HOLLYER, S. E. 1972. The sand and gravel resources of the area south and west of Woodbridge, Suffolk: Description of 1:25 000 resource sheet TM24. *Rep. Inst. Geol. Sci.*, No. 72/9, 128pp.

ALLSOP, J. M. and JONES, C. M. 1981. A pre-Permian palaeogeological map of the East Midlands and East Anglia. *Trans. Leicester Lit. Philos. Soc.*, Vol. 35, 28–33.

AMBROSE, J. D. 1973. The sand and gravel resources of the country around Layer Breton and Tolleshunt D'Arcy, Essex. Description of 1:25 000 resource sheet TL91 and part of TL90. *Rep. Inst. Geol. Sci.*, No. 73/8, 34 pp.

— 1974. The sand and gravel resources of the country west of Colchester, Essex: Description of 1:25 000 resource sheet TL92. *Rep. Inst. Geol. Sci.*, No. 74/6, 68 pp.

— 1975. The sand and gravel resources of the country east of Colchester, Essex: Description of 1:25 000 resource sheet TM02. *Miner. Assess. Rep. Inst. Geol. Sci.*, No. 14, 96 pp.

ANDERSON, E. M. 1942. *The dynamics of faulting.* (Edinburgh: Oliver and Boyd.)

ANDREWS, R. W. 1963. *Laminated Pleistocene deposits at Marks Tey, Essex.* Unpublished MSc thesis, Faculty of Science, University of London.

BADEN-POWELL, D. F. W. 1948. The Chalky Boulder Clays of Norfolk and Suffolk. *Geol. Mag.*, Vol. 85, 279–296.

BAKER, C. A. and JONES, D. K. C. 1980. Glaciation of the London Basin and its influence on the drainage pattern: a review and appraisal. Chapter 6 in *The Shaping of Southern England.* D. K. C. JONES (Editor). (Academic Press.)

BENTOR, Y. K. and KASTNER, M. 1965. Notes on the mineralogy and origin of glauconite. *J. Sediment. Petrol.*, Vol. 35, 155–166.

BERGGREN, W. A. 1971. Tertiary boundaries and correlations. Pp. 693–809 in *The micropalaeontology of the oceans.* B. M. FUNNELL and W. R. RIEDEL (Editors). (Cambridge: University Press.) 828 pp.

BISCAYE, P. E. 1965. Mineralogy and sedimentation of Recent deep sea clay in the Atlantic Ocean and adjacent seas and oceans. *Bull. Geol. Soc. Am.*, Vol. 76, 803–832.

BOOTH, S. J. and MERRITT, J. W. 1982. The sand and gravel resources of the country around Coggeshall, Essex. Description of 1:25 000 resource sheet TL82. *Miner. Assess. Rep. Inst. Geol. Sci.*, No. 102, 95 pp.

BOSWELL, P. G. H. 1929. The geology of the country around Sudbury (Suffolk). *Mem. Geol. Surv. G.B.*

BOULTON, G. S. and PAUL, M. A. 1976. The influence of genetic processes on some geotechnical properties of glacial till. *Q. J. Eng. Geol.*, Vol. 9, 159–194.

BRISTOW, C. R. 1985. The geology of the country around Chelmsford. *Mem. Br. Geol. Surv.*, Sheet 241, 108 pp.

— and COX, F. C. 1973. The Gipping Till: a reappraisal of East Anglian glacial stratigraphy. *Q. J. Geol. Soc. London*, Vol. 129, 1–37.

BROWN, G., CATT, J. A. and WEIR, A. H. 1969. Zeolites of the clinoptilolite-heulandite type in sediments of south-east England. *Mineral. Mag.*, Vol. 37, 480–488.

BROWN, J. 1843. On some Pleistocene deposits near Copford, Essex. *Proc. Geol. Assoc.*, Vol. 4, 164–165.

— 1852. On the Upper Tertiaries at Copford, Essex. *Q. J. Geol. Soc. London*, Vol. 8, 184–193.

CHANDLER, M. E. J. 1961. The Lower Tertiary floras of southern England. *Br. Mus. (Nat. Hist.) London.*

CLARKE, M. R. and AMBROSE, J. D. 1975. The sand and gravel resources of the country around Braintree, Essex:

Description of 1:25 000 resource sheet TL72. *Miner. Assess. Rep. Inst. Geol. Sci.*, No. 16, 111 pp.

CLAYTON, K. M. 1957. Some aspects of the glacial deposits of Essex. *Proc. Geol. Assoc.*, Vol. 68, 1–21.

COOPER, J. 1974. Report of field meeting to High Ongar, Essex. *Tertiary Times*, Vol. 2, 18–22.

— 1976. British Tertiary stratigraphical and rock terms formal and informal, additional to Curry, 1958, Lexique stratigraphique International. *Spec. Pap. Tertiary Res.*, No. 1.

CURRY, D. 1958. Lexique Stratigraphique International, Vol. 1, Part 3a XII, Great Britain Palaeogene. *Centre Nat. Recherche Scientifique.* (Paris.)

— 1965. The Palaeogene Beds of S.E. England. *Proc. Geol. Assoc.*, Vol. 76, 151–173.

— ADAMS, C. G., BOULTER, M. C., DILLEY, F. C., EAMES, F. E., FUNNELL, B. M. and WELLS, M. K. 1978. A correlation of Tertiary rocks in the British Isles. *Spec. Rep. Geol. Soc. London,*, No. 12, 72 pp.

DALTON, W. H. 1880. The geology of the neighbourhood of Colchester. *Mem. Geol. Surv. G.B.*

DAVIS, A. G. and ELLIOTT, G. F. 1957. The palaeogeography of the London Clay sea. *Proc. Geol. Assoc.*, Vol. 68, 255–277.

DAVISON, C. 1924. *A history of British Earthquakes.* (Cambridge.)

DINES, H. G. and EDMUNDS, F. H. 1925. The geology of the country around Romford. *Mem. Geol. Surv. G.B.*

DIXON, R. G. 1977. The palaeoecology of the Red Crag (Lower Pleistocene) of East Anglia. In X *Inqua Congress, Birmingham 1977, Volume of Abstracts.*

EASTON, C. H. 1973. The sand and gravel resources of the country around Terling, Essex. Description of 1:25 000 resource sheet TL71. *Rep. Inst. Geol. Sci.*, No. 73/5, 120 pp.

ELLISON, R. A. 1976. Great Cornard and Wormingford Mere Boreholes. *In IGS Boreholes 1975. Rep. Inst. Geol. Sci.*, No. 76/10.

— LAKE, R. D. and MOORLOCK, B. S. P. 1977. Bures and Great Bardfield boreholes. *In IGS Boreholes 1976. Rep. Inst. Geol. Sci.*, No. 77/10.

FISHER, O. 1863. On the brick-pit at Lexden, near Colchester. With notes on the Coleoptera by T.V. Wollaston. *Q. J. Geol. Soc. London*, Vol. 19, 393–401.

FITCH, F. J., HOOKER, P. J., MILLER, J. A. and BRERETON, N. R. 1978. Glauconite dating of Palaeocene–Eocene rocks from East Kent and the time-scale of Palaeogene volcanism in the North Atlantic region. *J. Geol. Soc. London*, Vol. 135, 499–512.

FRENCH, J. 1891. On the occurrence of Westleton Beds in part of north-western Essex. *Essex Nat.*, Vol. 5, 210–218.

FUNNELL, B. M. and WEST, R. G. 1977. Preglacial Pleistocene deposits of East Anglia. In *British Quaternary Studies, Recent advances.* F. W. SHOTTON (Editor). (Oxford: Clarendon Press.)

HAGGARD, H. J. E. 1972. The sand and gravel resources of the country around Witham, Essex. Description of 1:25 000 resource sheet TL81. *Rep. Inst. Geol. Sci.*, No. 72/6.

HARMER, F. W. 1900. The Pliocene deposits of the east of England. II, The Crag of Essex (Waltonian) and its relation to that of Norfolk and Suffolk. *Q. J. Geol. Soc. London*, Vol. 56, 705–738.

— 1904. The Great Eastern Glacier. *Geol. Mag.*, Vol. 41, 509–510.

HARRISON, R. K. and MERRIMAN, R. J. 1978. The mineralogy of the variegated clays of Reading Beds age, Knowl Hill, near Maidenhead. Guide to field excursion E6. *6th Int. Clay Conf.*, 5 – 8.

HESTER, S. W. 1965. Stratigraphy and palaeogeography of the Woolwich and Reading Beds. *Bull. Geol. Surv. G.B.*, No. 23, 117 – 123.

HEY, R., W. 1965. Highly quartzose pebble gravels in the London Basin. *Proc. Geol. Assoc.*, Vol. 76, 403 – 420.

— 1967. The Westleton Beds reconsidered. *Proc. Geol. Assoc.*, Vol. 78, 427 – 445.

— 1973. Alphamstone pit *in* ROSE, J. and TURNER, C. (Editors). Easter Field Meeting, 1973, Clacton. *Quat. Res. Assoc.*

— 1980. Equivalents of the Westland Green Gravels in Essex and East Anglia. *Proc. Geol. Assoc.*, Vol. 91, 279 – 280.

— and BRENCHLEY, P. J. 1977. Volcanic pebbles from Pleistocene gravels in Norfolk and Essex. *Geol. Mag.*, Vol. 114, 219 – 225.

HILDRETH, P. N. 1972. Upper Cretaceous and Lower Tertiary sequences in boreholes near Epping, Essex. *Bull. Geol. Surv. G.B.*, No. 38, 67 – 75.

HOLLYER, S. E. 1974. The sand and gravel resources of the country around Tattingstone, Suffolk: Description of the 1:25 000 resource sheet TM13. *Rep. Inst. Geol. Sci.*, No. 74/9, 86 pp.

HOLMES, T. V. 1892. The new railway from Grays Thurrock to Romford. Sections between Upminster and Romford. *Q. J. Geol. Soc. London*, Vol. 48, 365 – 372.

HOPSON, P. M. 1981. The sand and gravel resources of the country around Nayland, Suffolk. Description of 1:25 000 resource sheet TL93. *Miner. Assess. Rep. Inst. Geol. Sci.*, No. 85.

KING, C. 1970. The biostratigraphy of the London Clay in the London district. *Tertiary Times*, Vol. 1, 13 – 15.

— 1981. The stratigraphy of the London Clay and associated deposits. *Spec. Pap. Tertiary Res.*, No. 6, 158 pp.

KNOX, R. W. O'B. 1979. Igneous grains associated with zeolites in the Thanet Beds of Pegwell Bay, north-east Kent. *Proc. Geol. Assoc.*, Vol. 90, 55 – 60.

— and ELLISON, R. A. 1979. A Lower Eocene ash sequence in SE England, *J. Geol. Soc. London*, Vol. 136, 251 – 253.

— and HARLAND, R. 1979. Stratigraphical relationships of the early Palaeogene ash-series of NW Europe. *J. Geol. Soc. London.*, Vol. 136, 463 – 470.

— HARLAND, R. and KING, C. 1983. Dinoflagellate cyst analysis of the basal London Clay of southern England. *Newsl. Stratig.*, Vol. 12, 71 – 74.

KUKLA, G. J. 1977. Pleistocene land-sea correlations. I. Europe. *Earth Sci. Rev.*, Vol. 13, 307 – 374.

LAKE, R. D., ELLISON, R. A., HENSON, M. R. and CONWAY, B. W. 1986. The geology of the country around Southend-on-Sea. *Mem. Br. Geol. Surv.*

LISTER, T. R. 1971. In *Annu. Rept. Inst. Geol. Sci. for 1969*, p. 93.

MARKS, R. J. 1980. The sand and gravel resources of the country between Hatfield Heath and Great Waltham, Essex. Description of 1:25 000 resource sheets TL51 and TL61. *Miner. Assess. Rep. Inst. Geol. Sci.*, No. 52.

— and MERRITT, J. W. 1981. The sand and gravel resources of the country north-west of Halstead, Essex. Description of 1:25 000 resource sheet TL83. *Miner. Assess. Rep. Inst. Geol. Sci.*, No. 68.

— and MURRAY, D. W. 1981. The sand and gravel resources of the country around Sible Hedingham, Essex. Description of the 1:25 000 resource sheet TL73. *Miner. Assess. Rep. Inst. Geol. Sci.*, No. 82.

MIDDLEMISS, F. A. 1956. Field meeting at Brentwood and South Weald, Essex. *Proc. Geol. Assoc.*, Vol. 66, 317 – 319.

MILLWARD, D., ELLISON, R. A., LAKE, R. D. and MOORLOCK, B. S. P. (in press). The geology of the country around Epping. *Mem. Br. Geol. Surv.*

MITCHELL, G. F. 1977. Raised beaches and sea-levels. In *British Quaternary Studies, Recent advances:* F. W. SHOTTON (Editor), (Oxford: Clarendon Press.)

— PENNY, L. F., SHOTTON, F. W. and WEST, R. G. 1973. A correlation of Quaternary deposits in the British Isles. *Spec. Rep. Geol. Soc. London*, No. 4, 99 pp.

MONKTON, H. W. 1890. On the boulder clay in Essex. *Essex Nat.*, Vol. 4, 199 – 201.

ODIN, G. S., CURRY, D. and HUNZIKER, J. C. 1978. Radiometric dates from NW European glauconites and the Palaeogene time-scale. *J. Geol. Soc. London*, Vol. 135. 481 – 497.

OWEN, H. G. 1971. The stratigraphy of the Gault in the Thames Estuary and its bearing on the Mesozoic tectonic history of the area. *Proc. Geol. Assoc.*, Vol. 82, 187 – 207.

PERRIN, R. M. S., DAVIES, H. and FYSH, M. D. 1973. Lithology of the Chalky Boulder Clay. *Nature, Phys. Sci.*, Vol. 245, 101 – 104.

— ROSE, J. and DAVIES, H. 1979. The distribution variation and origins of pre-Devensian tills in eastern England. *Philos. Trans. R. Soc. London*, Series B. Vol. 287, 535 – 570.

PIKE, K. and GODWIN, H. 1953. The interglacial at Clacton-on Sea, Essex. *Q. J. Geol. Soc. London*, Vol. 108, 262 – 272.

PRESTWICH, J. 1847. On the main points of structure and probable age of the Bagshot Sands, etc. *Q. J. Geol. Soc. London*, Vol. 3, 378 – 409.

— 1850. On the structure of the strata between the London Clay and the Chalk, etc. Part i. Basement-Beds of the London Clay. *Q. J. Geol. Soc. London*, Vol. 6, 252 – 281.

— 1852. On the structure of the strata between the London Clay and the Chalk, etc. Part iii. The Thanet Sands. *Q. J. Geol. Soc. London*, Vol, 8, 253 – 264.

—1854. On the structure of the strata between the London Clay and the Chalk, etc. Part ii. The Woolwich and Reading Series. *Q. J. Geol. Soc. London*, Vol. 10, 75 – 170.

— 1890. On the relation of the Westleton Beds and Pebbly Sands of Suffolk, to those of Norfolk and on their extension inland. *Q. J. Geol. Soc. London*, Vol. 46, 84 – 181.

RHYS, G. H. (Compiler). 1974. A proposed standard lithostratigraphic nomenclature for the southern North Sea and an outline structural nomenclature for the whole of the (UK) North Sea. *Rep. Inst. Geol. Sci.*, No. 74/8, 14 pp.

ROSE, J., ALLEN, P. and HEY, R. W. 1976. Middle Pleistocene stratigraphy in southern East Anglia. *Nature, London*, Vol. 263, 492 – 494.

— and ALLEN, P. 1977. Middle Pleistocene stratigraphy in south-east Suffolk. *Q. J. Geol. Soc. London*, Vol. 133, 83 – 103.

— STURDY, R. G., ALLEN, P. and WHITEMAN, C. 1978. Middle Pleistocene sediments and palaeosols near Chelmsford. Essex. *Proc. Geol. Assoc.*, Vol. 89, 91 – 97.

— and TURNER, C. 1973. Easter Field Meeting, 1973, Clacton. *Quat. Res. Assoc.*

SALTER, A. E. 1905. On the superficial deposits of central and parts of southern England. *Proc. Geol. Assoc.*, Vol. 19, 1 – 50.

SAYER, A. R. and HARVEY, B. I. 1965. Records of Wells in the area of New Series One Inch (Geological) Great Dunmow (222) and Braintree (223) Sheets. *Water Supply Pap. Geol. Surv. G.B.*

SHACKLETON, N. J. 1977. Oxygen isotope statigraphy of the Middle Pleistocene. In *British Quaternary Studies, Recent advances.* F. W. SHOTTON (Editor). (Oxford: Oxford University Press.)

— and OPDYKE, N. D. 1973. Oxygen isotope and palaeomagnetic stratigraphy of Equatorial Pacific core V28 – 238: Oxygen isotope temperatures and ice volumes on a 10^5 year and 10^6 year scale, *Quat. Res.*, Vol. 3, 39 – 55.

SHOTTON, F. W. 1977. British dating work with radioactive isotopes. In *British Quaternary Studies, Recent advances.* F. W. SHOTTON (Editor). (Oxford: Oxford University Press.)

— SUTCLIFFE, A. J. and WEST, R. G. 1962. The fauna and flora from the brick pit at Lexden, Essex. *Essex. Nat.*, Vol. 31, 15–22.

SOLOMON, J. D. 1932. On the heavy mineral assemblages of the Great Chalky Boulder Clay. *Geol. Mag.*, Vol. 69, 314–320.

— 1935. The Westleton Series of East Anglia: its age, distribution and relations. *Q. J. Geol. Soc. London*, Vol. 91, 216–238.

STAMP, L. D. 1921. On cycles of sedimentation in the Eocene strata of the Anglo-Franco-Belgian Basin. *Geol. Mag.*, Vol. 58, 108–114, 146–157, 194–200.

STINTON, F. C. 1975. Fish otoliths from the English Eocene Part 1. *Palaeontogr. Soc.* [Monogr.], Vol. 129, No. 544, 1–56.

SZABO, B. J. and COLLINS, D. 1975. Ages of fossil bones from British Interglacial sites. *Nature, London*, Vol. 254, 680–682.

TRIPLEHORN, D. M. 1966. Morphology, internal structure and origin of glauconite pellets. *Sedimentology*, Vol. 6, 247–266.

TURNER, C. 1966. *Middle Pleistocene vegetational history and geology in East Anglia.* Unpublished PhD thesis, Queen's College, University of Cambridge.

— 1970. The Middle Pleistocene deposits at Marks Tey, Essex. *Philos. Trans. R. Soc. London*, Series B, Vol. 257, 373–437.

— 1975. The correlation and duration of the Middle Pleistocene interglacial periods in North-west Europe. Pp. 259–308 in *After the Australopithecines*, K. BUTZER and G. L. ISAAC (Editors).

WARD, D. J. 1978. The Lower London Tertiary (Palaeocene) succession of Herne Bay, Kent. *Rep. Inst. Geol. Sci.*, No. 78/10.

WARD, G. R. 1978. London Clay fossils from the M11 Motorway, Essex. *Tertiary Res.* Vol. 2, 17–21.

WEST, R. G. 1957. Interglacial deposits at Bobbitshole, Ipswich. *Philos. Trans. R. Soc. London*, Series B, Vol. 241, 1–31.

— 1963. Problems of the British Quaternary. *Proc. Geol. Assoc.*, Vol. 74, 147–186.

— and DONNER, J. J. 1956. The glaciations of East Anglia and the East Midlands: a differentiation based on stone-orientation measurements of the tills. *Q. J. Geol. Soc. London*, Vol. 112, 69–91.

WHITAKER, W. 1866. On the Lower London Tertiaries of Kent. *Q. J. Geol. Soc. London*, Vol. 22, 404–435.

— 1872. The geology of the London Basin. Vol. 4, Part 1. *Mem. Geol. Surv. G.B.*

— 1889. The geology of London, Vol. 1. *Mem. Geol. Surv. G.B.*

— PENNING, W. H., DALTON, W. H., BENNETT, F. J. 1878. The geology of the NW part of Essex and the NE part of Herts with parts of Cambridgeshire and Suffolk, (Explanation of sheet 47). *Mem. Geol. Surv. G.B.*

— and THRESH, J. C. 1916. The water supply of Essex from underground sources. *Mem. Geol. Surv. G.B.*

WHITE, H. J. O. 1932. The geology of the country around Saffron Walden. *Mem. Geol. Surv. G.B.*

WOOD, S. V. 1867. On the structure of the post-glacial deposits of the south-east of England. *Q. J. Geol. Soc. London*, Vol. 23, 394–417.

WOODLAND, A. W. 1970. The buried tunnel-valleys of East Anglia. *Proc. Yorkshire Geol. Soc.*, Vol. 37, 521–578.

WOOLDRIDGE, S. W. 1923. The minor structures of the London Basin. *Proc. Geol. Asscoc.*, Vol. 34, 175–193.

— 1926. The structural evolution of the London Basin. *Proc. Geol. Assoc.*, Vol. 37, 162–196.

— and HENDERSON, H. C. K. 1965. Some aspects of the physiography of the eastern part of the London Basin. *Trans. Inst. Br. Geogr.*, Vol. 21, 19–31.

WORSSAM, B. C. 1973. A new look at river capture and the denudation history of the Weald. *Rep. Inst. Geol. Sci.*, No. 73/17, 21 pp.

— and TAYLOR, J. H. 1969. The geology of the country around Cambridge. *Mem. Geol. Surv. G.B.*

WRIGLEY, A. 1924. Faunal divisions of the London Clay illustrated by some exposures near London. *Proc. Geol. Assoc.*, Vol. 35, 245–259.

— 1940. The faunal succession in the London Clay illustrated in some new exposures near London. *Proc. Geol. Assoc.*, Vol. 51, 230–245.

APPENDIX 1

Selected borehole logs

Wormingford Mere Borehole: BGS Ref TL 93 SW1; [TL 9267 3262]; surface level 26.5 m OD.

	Thickness m	Depth m
Quaternary		
Landslip: mottled orange brown silty clay derived from London Clay, with soft, wet peaty clay from 4.50 to 4.70 m	4.70	4.70
London Clay		
Bioturbated grey fine-grained sandy clay and clayey silt; black, rounded flint pebbles in the basal 5 cm	7.45	12.15
Woolwich and Reading Beds		
Interbedded mottled purplish clay with race, and greenish grey silt; glauconitic from 21.00 to 21.50 m	9.35	21.50
Thanet Beds		
Bioturbated grey fine-grained sand with occasional beds of glauconitic mottled clay, particularly from 32.0 to 33.0 m	11.50	33.00
Upper Chalk	seen to 0.50	33.50

Bures Borehole: BGS Ref TL 93 SW2; [TL 9120 3399]; surface level 30 m OD.

	Thickness m	Depth m
London Clay		
Clay, slightly fine-grained, sandy and silty, brown, stiff	0.60	0.60
Clay, as above, laminated; and slightly micaceous	0.80	1.40
Clay, pale brown, finely laminated with mica flakes	0.10	1.50
Silt, hard, partially iron-cemented. Well jointed	0.15	1.65
Clay, fine-grained, sandy and silty, orange mottled with patches of fine-grained micaceous sand. Original lamination visible in parts	0.45	2.10
Clay, grey brown and orange, finely laminated, occasional fine-grained micaceous sand streaks, laminae and patches. Becoming more silty and sandy with depth. Small pale orange siltstone fragments (up to 1 cm) at 3.00 m. Gradational base	0.90	3.00
Silt, clayey and fine-grained, sandy with fine-grained sand patches; micaceous. Slightly greenish tinge to dark sand grains (?glauconite pellets) at 3.40 to 3.50 m. Hard iron-stained siltstone at 3.60 m	0.60	3.60
Clay, fine-grained, sandy, orange-brown, laminated and micaceous	0.70	4.30

	Thickness m	Depth m
Septarian nodule: dark and pale grey siltstone showing bioturbation. *Core loss below the nodule*	0.40	4.70
Clay, fine-grained, sandy, mottled orange with a greenish tinge (?glauconite) in places. Micaceous and bioturbated	0.40	5.10
Clay, fine-grained, sandy, finely interbedded with silty clay and clayey fine-grained sand. The proportion of fine-grained sand increases with depth	1.90	7.00
Sand, fine-grained, clayey, orange-brown. Bioturbated but with some original lamination. Burrows infilled with fine-grained sand. One vertical burrow 3 mm across and 30 mm long	0.80	7.80
Sand, fine-grained, silty, orange-brown laminated and partially bioturbated in places. Interbedded grey silty clay to clayey silt	1.60	9.40
Sand, fine-grained, silty and silty clay, micaceous, bioturbated, olive-grey. Gradational base	1.10	10.50
Sand, fine-grained and silty, olive grey, bioturbated. Occasional clayey lenses. Pyrite nodules up to 10 mm diameter	0.90	11.40
Core loss	3.90	15.50
Woolwich and Reading Beds		
Clay, sandy; hard, bright blue-green, glauconitic	0.30	15.80
Silt, yellow-brown and bluish brown, laminated	0.20	16.00
Clay, stiff olive-brown and blue mottled, with race	3.00	19.00
Sand, fine- to medium-grained, clayey. Glauconitic, bright green. Clay content decreases with depth	1.50	20.50
Sand, fine to medium (as above). Bright green but with orange mottling. Bioturbated	0.50	21.00
Sand, fine- to medium-grained, alternating laminae of bright green and pale green to buff. Normally 1 to 2 per mm with apparently constant grain size throughout	1.30	22.30
Sand, fine- to medium-grained, laminated as above but dark green and pale green to buff laminae	0.90	23.40
Sand, fine-grained, greenish grey with glauconite pellets and scattered mica flakes, bioturbated; with buff, micaceous, fine-grained sand patches. Becoming more clayey at 24.00 to 24.10 m. Occasional patches of dark bottle green glauconitic sand down to 24.60 m. Sharp base with termination of glauconite content	3.40	26.80
Thanet Beds		
Sand, fine-grained as above, medium grey, bioturbated and micaceous	seen to 7.20	34.00
Core loss between about 27.50 and 33.50 m		

Marks Tey No. 1 Borehole: BGS Ref TL 92 SW88; [TL 9066 2473]; surface level 32.1 m OD.

	Thickness m	Depth m
Head		
Silt, orange brown and grey mottled with occasional angular flint gravel up to 3 cm diameter. Becoming increasingly gravelly towards the base	1.50	1.50
Sand and gravel up to 3 cm diameter; mostly angular to subangular	0.50	2.00
Lacustrine Deposits		
Silt, clayey, bluish grey, weathering quickly to greenish grey; slightly micaceous and with race. Slightly more clayey from 2.20 m. Greenish brown mottling and with abundant race up to 2 cm diameter from 2.64 to 2.80 m	0.80	2.80
Core loss	0.15	2.95
Clay, silty; soft greyish green. Organic streaks and occasional bands containing medium-grained sand with chalk and/or race pellets	1.00	3.95
Core loss	0.23	4.18
Silt, clayey; soft dark grey-green. Small chalk fragments at 4.38 to 4.50 m. Gradational base with increasing clay content	0.32	4.50
Clay, silty, with occasional fine- to medium-grained sand beds with small race and/or chalk pellets up to 2 mm diameter from 4.95 m. Some reworking with buff silt streaks. Gradational base	0.64	5.14
Silt, clayey; greenish grey with clasts of reworked dark grey organic silt up to 3 mm diameter. Sand beds with race or chalk as above. Homogeneous throughout giving a crumbly texture on freshly broken surfaces. Feint lamination seen between 5.14 and 5.20 m. Becoming more clayey at 5.60 to 6.68 m with pale olive mottling. Moss fragments seen on bedding plane inclined 5° to the horizontal, at 5.90 m	1.54	6.68
Clay, silty; mottled dark grey and pale greenish grey. Occasional shell fragments; friable; plant remains and vivianite in places	0.62	7.30
Clay, silty, with fine sand grains in places; mottled grey with clasts of greyish green clayey silt; slightly micaceous and with race in places. Gradational base	0.70	8.00
Silt, clayey; mottled pale and dark olive giving a speckled appearance resulting from fine brecciation. Slightly micaceous	0.30	8.30
Core loss	0.30	8.60
Silt, clayey to silty clay, pale greenish grey. Becoming speckled with small clasts of dark grey clayey silt (up to 2 mm diameter) from 8.70 to 8.75 m. Gradational base	0.15	8.75
Silt, clayey to silty clay; mottled pale greyish green and dark grey. Finely brecciated. Scattered shells from 9.55 to 9.95 m; slightly micaceous	1.19	9.94
Silt, mottled pale greenish grey and grey-green; friable and finely brecciated; scattered shells, organic material. Becoming clayey at 10.50 m. Disturbed bedding results in core fracture at 45°	1.06	11.00

	Thickness m	Depth m
Silt, clayey and very silty clay; stiff; brecciated; pale grey-green and medium grey mottling. Shell fragments and organic debris scattered throughout, with leaf impressions on some relict bedding surfaces. Discrete clasts up to 1 cm diameter, increasing to 3 cm and containing shell debris from 12.30 to 12.90 m. Becoming dark grey with pale greyish green mottling from 12.10 to 12.90 m	1.90	12.90
Core loss	0.20	13.10
Silt, organic, laminated, medium brown	0.10	13.20
Core loss	0.20	13.40
Silt, organic, pale brown	0.15	13.55
Core loss	0.18	13.73
Organic material, detrital, fissile (like leaf mould layers). Sharp base with no roots	0.27	14.00
Silt, friable, pale grey-green, with shell debris. Possible faint bedding trace. ?Gradational base	0.40	14.40
Silt, slightly clayey and with fine-grained sand patches; massive	0.20	14.60
Silt, organic in the top 10 cm; friable, micro-brecciated	0.30	14.90
Core loss	0.40	15.30
Silt, clayey, mottled dark grey and pale olive-brown, slightly micaceous, friable, micro-brecciated	0.30	15.60
Silt, clayey, pale greyish green; laminated with occasional streaks of pale buff silt	0.80	16.40
Core loss	0.35	16.75
Silt, with organic detritus, dark brown; laminated with abundant carbonised plant remains from 17.30 to 18.10 m. Occasional thin partings of pale buff silt.	1.75	18.50
Silt, organic, dark brown to black, laminated. Shell fragments from 19.50 to 20.00 m	1.50	20.00
Silt, hard, leathery and fissile with low density; brown, becoming pale orange brown with shell debris at 20.10 m	5.00	25.00
Silt, greenish grey and pale buff in alternating laminae, with 2 to 4 per mm (varved sequence); soft to firm. Very finely laminated from about 30.00 m (10 per mm). Local beds with no pale buff laminae	8.00	33.00
Silt, clayey, slightly greenish grey; laminated and more fissile than above with alternating grey and pale buff silt; micaceous	1.40	34.40
Silt, clayey; softer than above and more finely laminated with alternating greyish green and buff laminae (5 to 10 per mm)	0.15	34.65
Silt, as above; less finely laminated and with increasing dark grey organic laminae; Some shell fragments from 35.00 to 35.15 m	0.50	35.15
Clay, silty; mottled dark grey, black and pale olive; shelly in places and with reed stems common on lamination planes	0.85	36.00
Clay, dark grey with chalk, flint and quartzite pebbles; less chalky in places and with organic material down to 36.40 m	0.40	36.40

	Thickness m	Depth m
Chalky Boulder Clay, with rounded chalk, angular and rounded flint pebbles	seen to 7.30	43.70

Marks Tey No. 2 Borehole: BGS Ref TL 92 SW89; [TL 9161 2440]; surface level 28.8 m OD.

	Thickness m	Depth m
Head		
Gravel, angular to subrounded with coarse-grained sand; slightly clayey	0.50	0.50
Boulder Clay		
Clay, gravelly and sandy; orange with blue veining and race nodules	1.00	1.50
Clay, olive-grey with pebbles and chalk clasts	18.20	19.70
Glacial Silt		
Clay, fine sandy and silty to clayey silt; olive and brownish olive alternating laminae (4 to 5 per mm) and local buff fine-grained sand laminae from 21.10 m. Scattered dark grey patches (bioturbation?) throughout and common organic dark grey laminae from 22.00 m	4.30	24.00
Glacial Sand and Gravel		
Gravel, fine- to coarse-grained, chalky and sandy. Coarse sandy clay with mainly rounded flints and chalk pebbles from 28.5 to 28.8 m	seen to 6.00	30.00

Marks Tey No. 3 Borehole: BGS Ref TL 92 SW90; [TL 9343 2438]; surface level 25.2 m OD.

	Thickness m	Depth m
Lacustrine Deposits		
Sand, clayey and silty with some gravel, mainly angular and subrounded flints	2.00	2.00
Silt, clayey, pale grey and organic, dark grey; and orange; finely laminated. Becoming mainly pale grey calcareous silt. Sharp base	1.23	3.23
Mudstone, black, organic, fissile, highly fissured. Sharp base. Interbed of silt: pale greenish grey, laminated, with common gastropods from 3.52 to 3.55 m	0.37	3.60
Core loss	0.60	4.20
Silt, pale greenish grey and dark grey, finely laminated. Gradational base	0.05	4.25
Clay, medium grey, with dark grey organic wisps and shell fragments; finely laminated. Few organic wisps between 4.50 to 5.00 m. Abrupt change at base (junction not seen)	2.35	6.60
Clay, very silty, olive grey, finely laminated	0.10	6.70
Silt, clayey, mottled grey and orange	0.30	7.00
Silt, clayey, medium grey, laminated; with occasional shells. Scattered chalk pellets up to 2 mm diameter at around 8.70 m	3.70	10.70
Silty clay to silt with some fine sand; dark grey colour; firm to stiff; micaceous; laminated to very well laminated 11.70 to 12.30 m; some small shell fragments	1.80	12.50

	Thickness m	Depth m
(*Core lost 12.50 to 14.00 m*—base of grey laminated clay at *c*.13.50 m)	1.50	14.00
Sand, fine to medium grade, grey, silty with small chalk fragments	0.30	14.30
Boulder Clay		
Clay, very sandy; angular flints concentrated at 14.70 to 14.80 m passing downwards into:	0.50	14.80
Chalky boulder clay, sandy, dark grey-brown, small chalk and flint fragments	8.20	23.00
Sand, clayey, fine to medium grade; brown to greenish brown; occasional angular flints	0.10	23.10
Clay, slightly sandy, flinty, with rare rounded quartz pebbles; dark grey-brown colour	0.85	23.95
(*Core lost 24.00 to 24.85 m*)	*c.*0.55	*c.*24.50
Glacial Silt		
Clay, dark grey to greenish grey with pale grey and black streaks; silty, fine sandy, micaceous; well laminated in places with laminae (1 to 2 mm) of grey and black silt; occasional small rounded pebbles (c.1 cm)	1.10	25.60
Glacial Sand and Gravel		
Clayey, medium grade sand and subangular to rounded flint gravel becoming finer with depth (angular flints 0.5 to 1 cm)	seen to 3.40	29.00

Marks Tey No. 4 Borehole: BGS Ref TL 92 SW91; [TL 9242 2471]; surface level 32.6 m OD

	Thickness m	Depth m
Sand and Gravel (?glacial in part)		
Sandy top soil	0.20	0.20
Sand, medium grade, clayey with small angular and rounded flints, orange	1.80	2.00
Sand, fine, clay-free	1.10	3.10
Sand, fine, chalky	0.10	3.20
Sand, fine to medium, clay-free	0.80	4.00
Sand, fine to medium, silty in places, laminated	0.20	4.20
Gravel, fine grade, angular	0.10	4.30
Sand, fine, bedded	0.50	4.80
Sand, fine, clay-free, structureless	4.10	8.90
Sand, fine to medium with occasional gravel—angular flints 4 cm; small fragments of chalk and flint in places; grey	13.10	22.00
Clay, silty, micaceous, laminated; grey	0.50	22.50
Gravel, fine grade; coarse sandy matrix; chalky; rounded chalk and angular flint	1.00	23.50
Gravel, fine grade; coarse sandy matrix; chalky; local angular flints	seen to 9.50	33.00

The logs of boreholes AA, BB and GG are taken verbatim from C. Turner, 1970. The Middle Pleistocene Deposits of Marks Tey, Essex. *Philos. Trans. R. Soc. London*, (B) 257, 373–437.

Borehole AA per C. Turner; [TL 9126 2441]; surface level +32.6 m OD

	Thickness m	Depth m
Top soil	0.40	0.40
Yellow-brown silty sand, greyer and more clayey below	1.90	2.30

	Thickness m	Depth m
Stiff grey, finely silty clay	0.60	2.90
Grey-brown silty clay mud	0.75	3.65
Blackish brown, medium–fine detritus mud	0.40	4.05
Brecciated, light brown shelly clay mud	0.30	4.35
Dark brown shelly detritus mud with a few layers of brecciated clay mud	0.40	4.75
Grey-brown to grey silty clay mud, sometimes shelly and with occasional small stones	2.00	6.75
Grey chalk sand, coarser at the base	0.40	7.15
Bluish grey chalky boulder clay —	touched	—

Borehole BB per C. Turner; [TL 9091 2438]; surface level + c.32.3 m OD

	Thickness m	Depth m
Top soil	0.30	0.30
Mottled brown and grey clay with calcium carbonate concretions	1.45	1.75
Silty grey-brown clay mud	2.25	4.00
Dark brown, medium–fine detritus mud with some bands of clay mud, becoming shelly below 5 m	1.25	5.25
Brecciated grey-brown clay mud, occasionally with thin layers of detritus mud	1.00	6.25
Grey clayey sand and silt; boring ended in coarse sand at 6.85 m	0.60	6.85

Borehole GG per C. Turner; [TL 9108 2443]; surface level 15.9 m OD

	Thickness m	Depth m
Mottled orange-grey clay	1.77	1.77
Laminated, medium to dark grey clay, with distinctive light clay layers at 1.86 to 1.94 and 2.54 m. Passing into	1.37	3.14
Unevenly banded clay, showing layers, 5 to 10 cm thick, of grey, orange-grey and grey-brown clay in irregular order which merge indistinctly into one another. This clay sometimes contains fine brecciated flakes and nodules of orange-grey clay and mud. Passing into	0.90	4.04
Distincly and finely laminated grey clay and grey-brown clay mud. Lighter coloured layers of grey silty clay, varying in thickness from 0.3–2 cm, alternate with grey-brown, brown or blackish layers of silt, 0.3–1 cm thick. Towards the base of the bed the darker bands become more pronounced and the lighter ones thinner. When broken across, the material of the darker layers reveals many platy reworked fragments of brown silty, sometimes shelly clay mud and of grey clay. The lighter layers also contain some reworked material	3.13	7.17

	Thickness m	Depth m
Khaki-brown clay mud (drying to grey), thoroughly brecciated into angular fragments 1 to 10 cm across, lying in a matrix of finer fragments of the same material. Most of these fragments show well defined lamination of the order of 1 to 2 mm with alternating dark and yellow buff laminae	0.54	7.71
Faintly laminated khaki-brown clay mud. Dip of the laminations about 50°; probably a single large inclined block of clay within the breccia	0.71	8.42
As 7.17 to 7.71 m, with lenses of dark grey silt between the brecciated blocks at 8.69 to 8.79 m and irregular fine clay lenses and a 1 cm dark organic silt layer at 11.16 to 1.18 m	4.09	12.51
Contorted seam of grey clay containing brecciated fragments of brown clay mud. (Sole of slumped mass of breccia)	0.03	12.54
Finely laminated, rather fissile, dark, khaki or grey-brown clay mud, becoming greyish in colour and very light on drying out. Fine, silty, organic, brown laminae alternate with grey, buff-yellow or white laminae; each lamination pair is generally less than 1 mm thick. A slight irregularity of the lamination occurs between 12.82 and 12.85 m and a thin brecciated horizon occurs 14.55 to 14.60 m. Below 17.07 m the lamination becomes finer and sometimes indistinct. Passing into	5.91	18.45
Laminated dark to khaki-brown and grey clay mud showing irregular alternating grey and brown bands, 0.5 to 3 cm thick. Lamination often poorly developed	0.35	18.80
Silty grey clay with a seam of dark grey sand and a 5 cm angular flint at 18.91 m	0.12	18.92
Banded and laminated silty clay mud, at 18.45 to 11.80 m, becoming a more even-coloured brownish grey laminated clay below 19.81 m. Passing smoothly into	1.18	20.10
Firm grey clay, faintly laminated, with occasional shell fragments between 20.12 and 20.22 m, thin lamellae of redeposited $CaCO_3$ from 20.40 to 20.70 m and fine inorganic black speckling	0.77	20.87
Sandy to silty grey clay with two thin seams of sand and fine chalk gravel at 20.95 and 21.16 m	0.33	21.20
Medium to coarse grey sand with fine chalk gravel, derived Cretaceous foraminifera and sponge spicules	0.21	21.41
Finely bedded, silty, grey clay. Artesian water was struck at this level. It rose to the surface and continued to flow at a rate of 14 000 to 18 000 litres per hour for a month before the borehole was sealed off	seen to 0.03	21.44

APPENDIX 2

Mineralogy of Wormingford Mere and Bures Boreholes

Table 4   Mineralogy of Wormingford Mere Borehole [TL 9267 3262]

Formation and Lithology	Depth (m)	XRD No. DX	Clay minerals* (with relative abundances)	Quartz %	Dolomite %	Calcite %	Feldspar	Other minerals detected	Registered No. in national collections
Land-slipped London Clay									
Yellow-brown silty clay	3.0-3.5	1838	illite > smectite > kaol,chlor	20	n.d.	n.d.	Na > K(w)		MR 34899
Yellow-brown silty clay	3.9-4.1	1839	illite > smectite > kaol,chlor	17	n.d.	n.d.	K > Na(w)		MR 34900
Ochreous silty clay	4.5-4.7	1840	illite > smectite > kaolinite	18	n.d.	n.d.	<1%	goethite	MR 34901
London Clay									
Micaceous clayey fine sand	8.7-9.5	1841	illite > smectite > kaol > chlor	33	n.d.	n.d.	Na > K(s)	gypsum, pyrite	MR 34902
Micaceous clayey fine sand	9.5-10.0	1842	illite > smectite ≥ kaol > chlor	30	<1	n.d.	Na = K(w)	gypsum	MR 34903
Micaceous clayey fine sand	10.0-11.0	1843	illite > smectite > kaol > chlor	28	<1	n.d.	K > Na(m)	gypsum	MR 34904
Clayey sand (+ clay lenses)	12.10-12.15	1844	illite > smectite > kaol > chlor	21	n.d.	n.d.	Na(w)	(E 52742) ilmenite (X 7969)	MR 34905
Woolwich and Reading Beds									
Orange, grey veined clay	14.3-14.5	1845	illite > kaol ≥ smectite > chlor	38	2	7	Na = K(w)	goethite	MR 34906
Yellow/brown clayey sand	15.3-15.7	1846	illite > smectite ≥ chlor. > kaol	55	<1	5	Na(m)		MR 34907
Brown, blue veined clay	16.2-17.1	1847	illite > smectite ≥ kaol > chlor	19	n.d.	4	<1%	goethite	MR 34908
Greyish khaki clayey sand	17.1-18.1	1848	smectite > illite	48	n.d.	3	n.d.		MR 34909
Khaki clayey sand	18.5-19.5	1849	smectite > illite	52	n.d.	n.d.	K(w)		MR 34910
Glauconitic sand	21.0-22.0	1850	smectite > illite (+ glauconite)	50	n.d.	n.d.	<1%	hematite, goethite	MR 34911
Thanet Beds									
Clayey, fine sand	23.0-26.5	1851	smectite > illite > chlorite	35	10	n.d.	Na, Ca > K(m)	clinoptilolite	MR 34912
Glauconitic clay	32.0-33.0	1852	smectite (+ glauconite)	12	n.d.	6	Ca > K(m)	clinoptilolite, gypsum (E 51375)	MR 34913

Table 5   Mineralogy of Bures Borehole [TL 9120 3399]

Formation and Lithology	Depth (m)	XRD No. DX	Clay minerals* (with relative abundances)	Quartz %	Dolomite %	Calcite %	Feldspar	Other minerals detected	Registered No. in national collections
London Clay									
Brown sandy clay	10.5	1823	illite > smectite > kaol,chlor	40	n.d.	n.d.	Na > K(s)		MR 34891
Brown sandy clay	11.1	1824	illite > smectite > kaol,chlor	27	<1	n.d.	K(m)	gypsum	MR 34892
Woolwich and Reading Beds									
Brown/blue mottled clay	17.0	1825	illite > smectite > kaol > chlor	12	n.d.	<1	<1%	goethite	MR 34893
Grey, reddish sandy clay	19.3	1826	smectite > illite	26	n.d.	n.d.	<1%	goethite	MR 34894
Grey/brown sandy clay	19.5	1827	smectite > illite	40	n.d.	n.d.	<1%		MR 34895
Glauconite sand	22.5	1828	smectite ≃ illite	30	n.d.	n.d.	Na(w)	goethite	MR 34896
Thanet Beds									
Micaceous clayey fine sand	26.7	1829	smectite > illite ≥ chlorite	52	6	n.d.	Na > Ca > K	clinoptilolite	MR 34897
Micaceous clayey fine sand	33.5	1830	smectite > illite ≥ chlorite	47	n.d.	n.d.	Na, Ca > K(m)	clinoptilolite, gypsum	MR 34898

*Chlorite is listed with 'Clay minerals' but occurs mostly as flakes >2 μm with only minor amounts of less than 1% in the <2 μm fractions.

≠ Feldspar symbols:
    K = microcline/orthoclase   (w) = weak peaks (1–2%)
    Na = albite-oligoclase   (m) = medium peaks (2–5%)
    Ca = andesine-labradorite   (s) = strong peaks (5–10%)
                                n.d.= not detected

APPENDIX 3

List of Geological Survey photographs

Copies of these photographs are deposited for reference in the British Geological Survey library, Keyworth, Nottingham NG12 5GG. Black and white prints and slides can be supplied at a fixed tariff, and in addition colour prints and transparencies are available for all the photographs. The photographs were taken by Messrs H. J. Evans and C. J. Jeffrey.

All numbers belong to Series A.

12364	Kesgrave Sands and Gravels overlying Crag. Ferriers Farm Pit
12365	Sub-glacial deposits overlying Kesgrave Sands and Gravels, gravel pit 2 km south of Sible Hedingham.
12500	Kesgrave Sands and Gravels, Shalford.
12501	Boulder Clay overlying Kesgrave Sands and Gravels, Beazley End.
12502	Upper part of sand and gravel sequence, Beazley End.
12503	ARC Stanway gravel pit.
12504	Glacial Sand and Gravel, Stanway.
12505	Kesgrave Sands and Gravels, Stanway.
12506	Kesgrave Sands and Gravels, Stanway.
12507	Drift deposits, Stanway.
12508	Ice wedge cast, Stanway.
12509	Kesgrave Sands and Gravels, Stanway.
12510	Climbing ripple lamination, Stanway.
12511	Glacial Sand and Gravel, Stanway.
12512	Detail of Kesgrave Sands and Gravels, Stanway.
12513	Drift deposits, Stanway.
12514	Stour valley, Nayland.
12515	Stour valley, Nayland.
12516	Stour valley, Wormingford.
12517	Kesgrave Sands and Gravels, Kilowen sand-pit.
12524	Kesgrave Sands and Gravels, Kilowen sand-pit.
12525	Head deposits, Kilowen sand-pit.
12526	Kilowen sand-pit.
12527	Glacial silt and chalky gravel, Castle Hedingham.
12528	Glacial silt beneath Head, Castle Hedingham.
12529	Head overlying glacial deposits, Castle Hedingham.
12530	Drift deposits, Foxborough pit.
12531	Chalky Boulder Clay, Foxborough pit.
12532	Kesgrave Sands and Gravels, Foxborough pit.
12533	Boulder Clay plateau, Burtons Green.
12534	Blackwater valley, Bradwell.
12535	Reclamation of Bradwell pit.
12536	Blackwater valley, Coggeshall.
12537	Colne valley, Chappel.
12538	Colne valley, Chappel.
12539	First Terrace of Colne valley, Earls Colne.
12540	Colne valley, Colne Engaine.
12542	Dissected boulder clay plateau, Great Maplestead.
12543	Colne valley, Great Maplestead.
12544	Colne valley, Sible Hedingham.
12545	Colne valley, Sible Hedingham.
12546	Great Tey church.
12547	Old barn with Boulder Clay daub, Sible Hedingham.
12555	Stour valley, Great Hickbush.
12556	Base of chalky boulder clay, Alphamstone.
12558	Glacial Sand and Gravel, Ferriers Farm.
12559	Glacial Sand and Gravel, Ferriers Farm.
12560	Convolutions beneath Boulder Clay, Ferriers Farm.
12561	Kesgrave Sands and Gravels, Ferriers Farm.
12562	Head overlying lacustrine deposits, Marks Tey.
12563	Upper part of lacustrine deposits, Marks Tey.
12564	Lacustrine deposits, Marks Tey.
12565	Detail of interglacial deposits, Marks Tey.
12566	Working face, Marks Tey brick-pit.
12557	Ferriers Farm gravel pit.
13005	London Clay injection structure, Stanway.
13006	Kesgrave Sands and Gravels overlain by Glacial Sand and Gravel, Stanway.
13007	Kesgrave Sands and Gravels, Stanway.

INDEX OF FOSSILS

GENERAL INDEX

BRITISH GEOLOGICAL SURVEY

Keyworth, Nottinghamshire NG12 5GG

Murchison House, West Mains Road,
Edinburgh EH9 3LA

The full range of Survey publications is available
through the Sales Desks at Keyworth and
Murchison House. Selected items are stocked by
the Geological Museum Bookshop, Exhibition
Road, London SW7 2DE; all other items may be
obtained through the BGS London Information
Office in the Geological Museum. All the books
are listed in HMSO's Sectional List 45. Maps are
listed in the BGS Map Catalogue and Ordnance
Survey's Trade Catalogue. They can be bought
from Ordnance Survey Agents as well as from
BGS.

*On 1 January 1984 the Institute of Geological Sciences
was renamed the British Geological Survey. It continues to
carry out the geological survey of Great Britain and
Northern Ireland (the latter as an agency service for the
government of Northern Ireland), and of the surrounding
continental shelf, as well as its basic research projects. It
also undertakes programmes of British technical aid in
geology in developing countries as arranged by the Overseas
Development Administration.*

*The British Geological Survey is a component body of the
Natural Environment Research Council.*

HER MAJESTY'S STATIONERY OFFICE

HMSO publications are available from:

HMSO Publications Centre
(Mail and telephone orders)
PO Box 276, London SW8 5DT
Telephone orders (01) 622 3316
General enquiries (01) 211 5656

HMSO Bookshops
49 High Holborn, London WC1V 6HB
  (01) 211 5656 (Counter service only)
258 Broad Street, Birmingham B1 2HE
  (021) 643 3757
Southey House, 33 Wine Street, Bristol BS1 2BQ
  (0272) 24306/24307
9 Princess Street, Manchester M60 8AS
  (061) 834 7201
80 Chichester Street, Belfast BT1 4JY
  (0232) 238451
13a Castle Street, Edinburgh EH2 3AR
  (031) 225 6333

HMSO's Accredited Agents
(see Yellow Pages)

And through good booksellers